Lecture Notes in Electrical Engineering

Volume 374

About this Series

"Lecture Notes in Electrical Engineering (LNEE)" is a book series which reports the latest research and developments in Electrical Engineering, namely:

- Communication, Networks, and Information Theory
- Computer Engineering
- Signal, Image, Speech and Information Processing
- Circuits and Systems
- Bioengineering

LNEE publishes authored monographs and contributed volumes which present cutting edge research information as well as new perspectives on classical fields, while maintaining Springer's high standards of academic excellence. Also considered for publication are lecture materials, proceedings, and other related materials of exceptionally high quality and interest. The subject matter should be original and timely, reporting the latest research and developments in all areas of electrical engineering.

The audience for the books in LNEE consists of advanced level students, researchers, and industry professionals working at the forefront of their fields. Much like Springer's other Lecture Notes series, LNEE will be distributed through Springer's print and electronic publishing channels.

More information about this series at http://www.springer.com/series/7818

M.V. Hariharan · S.D. Varwandkar
Pragati P. Gupta

Modular Load Flow for Restructured Power Systems

Springer

M.V. Hariharan
Department of Electrical Engineering
Indian Institute of Technology Bombay
Mumbai, Maharashtra
India

Pragati P. Gupta
Department of Electrical Engineering
Veermata Jijabai Technological Institute
Mumbai, Maharashtra
India

S.D. Varwandkar
Department of Electrical Engineering
Veermata Jijabai Technological Institute
Mumbai, Maharashtra
India

ISSN 1876-1100 ISSN 1876-1119 (electronic)
Lecture Notes in Electrical Engineering
ISBN 978-981-10-9177-3 ISBN 978-981-10-0497-1 (eBook)
DOI 10.1007/978-981-10-0497-1

Printed on acid-free paper

This Springer imprint is published by SpringerNature
The registered company is Springer Science+Business Media Singapore Pte Ltd.

To Gustav Robert Kirchhoff and Werner Karl Heisenberg whose insights opened for us a whole new perspective hitherto unexplored…

Preface

All truth passes through three stages. First, it is ridiculed; second, it is violently opposed; and the third, it is accepted as being self-evident...

Arthur Schopenhauer (1788–1860)

Having taught the subject of power systems to a large number of students over several years the authors felt that a re-look is necessary at some conventional methods of analysis. In this book, the authors have subjected the time-honoured load flow to a close scrutiny. Books on electrical circuits do not go beyond the usual electrical domain, i.e. all sources are voltage/current sources, or their variants. Power source does not find mention anywhere. *Scalar* power driving *phasor*-governed ac electrical circuits appears to be alien to prevalent thinking, but that is how it is! Turbine-driven synchronous generators perform precisely this role in power systems. Given the power source, how does one then find the voltages and currents spread all over the network? How does the power quantum of generator get distributed among network elements? We try to find answers to these interesting questions in this book. The analysis is an eye opener. We enter into a *wonderland* where all final entities, such as flows, losses, voltage magnitudes are real numbers, or quanta! Phasors that hitherto constituted the so-called state of the system do not appear either in the input or in the output. We thus discover a new load flow procedure—Modular Load Flow—which works with only scalars!

Historically power systems have had a structure of vertical integration. In a given geographical area, generation, transmission and distribution systems all belonged to one commercial agency. Connections to neighbouring systems belonging to another commercial agency were through tielines. In such a structure, it was easy to settle commercial issues like payments for power exchange between the two systems. With the increase in number of new generation and transmission facilities and with the increase in the number of stakeholders, concept of horizontal integration emerged about two decades ago. Multiple agencies pumped power into a large interconnected grid with transmission lines belonging to different agencies.

While this trend had the beneficial effect of competition and cost reduction, problems arose in determining the extent of use of a given transmission facility by a specific agency.

A restructured and deregulated system is all about injection of electrical powers—quanta—into the network where stakeholders may desire to know for commercial reasons, their individual contributions to flows, losses and voltages in the system. Modular load flow provides an effective tool for this purpose. To cap it all, the procedure is one-shot, a closed form solution! Ever since mid-1950s only the iterative load flow has been in vogue in almost all studies in power systems. It has had its own problems though, namely, non-convergence in ill-conditioned, or, heavily loaded systems, and multiple solutions. Continuation load flow and various other improvisations to ensure convergence are well-known workarounds. A fact that is not usually recognised is that the output of a conventional load flow is only a *possibility* and not a *solution* that would indicate reality. What if studies were conducted with a different slack bus? What if the voltage specifications at PV buses were different in different studies? There would be a different solution every time though one knows for sure that the generator injection pattern has not changed at all. Somewhere there appears to be a disconnect. It is as if the system operator expects generating stations to adjust reactive powers and voltages in sync with those used by him in the load flow and, to continuously keep him updated about reactive power limits and voltages. Switching of PV bus to PQ bus, when the limits are violated, crucially depends on this information. In deregulated systems, these expectations are not only far-fetched but are highly, if not totally, irrelevant. Modular load flow is different in this respect. Only information that is necessary is the values of injected real and reactive powers. These measurements are usually available in control centres. There are no issues of convergence. Solution is analytically exact. Errors, if any, occur due to minor approximation in passive circuit formulation or, in the data.

Kirchhoff state is a new concept introduced by the authors. It offers a new way of viewing the decoupling that naturally exists in power systems but does not seem to have been reported in the literature. The decoupling offers many advantages especially in deregulated systems. Outage analysis, economic optimization and voltage stability problems are posed in new formats. Authors have preferred to hand-hold student readers by giving numerical examples to illustrate every new concept or procedure when it is introduced.

Contingency analyses use distribution factors which are generally straightforward to calculate in centrally administered power systems. A global base case load flow is a prerequisite for this purpose. This cannot be reliably conducted in restructured multi-area systems. Modular load flow can deal with this situation easily. Security aspects of a part of Indian grid preceding the blackout on 30 and 31 July 2012 are discussed.

Deregulation has come to stay, and tracking of power flow is needed to segregate electrical contributions of various commercial agencies. System analysts have to find new tools to meet analytical requirement of such new developments. Analysis given in this book fulfils this need. We hope suggestions made in the book will find acceptance among power system teachers, graduate students and professionals interested in the analysis of restructured systems. It should be of help to regulators in resolving conflicts. There is an elegant theory based on orthogonality, Hilbert spaces and Dirac structures behind this analysis. Informal and brief discussion on this theory in Chap. 10 should entice avid researchers.

The unique analytical experiment, that is, modular load flow is presented in the form of lecture notes and illustrated with several examples. Only steady-state performance has been addressed in the book. Dynamic aspects are under research.

Acknowledgements

We owe gratitude to many individuals for their encouragement, support and co-operation in this work. Professor K.M. Kulkarni (VJTI) saw the promise in this concept while it was in nascent stage. The authors had discussions with Professor H. Narayanan, Indian Institute of Technology Bombay, Dr. M. Ramamoorty, former Professor, IIT Kanpur, former Director General, Central Power Research Institute and currently Chancellor, KL University, Andhra Pradesh, Prof. K.R. Padiyar and Prof. P.S. Nagendra Rao (via *inpowerg*) of Indian Institute of Science, Bangalore. These discussions have revealed interesting aspects of the theory. Professor S.A. Patekar, former Professor of Computer Engineering, VJTI, and currently, Principal at Vidyalankar Institute of Technology, Mumbai, took great interest in our work. We are grateful to them all for extending our horizons. Authors are thankful to Dr. Nathaniel Taylor, Researcher (who teaches too), KTH/ETK, Sweden, and Dr. Quiang Wei, Electric Power Research Institute of Henan Electric Power Company, Zhengzhau China; both reworked our calculations and commented on it in early stages. Application of this concept needed understanding of ground realities in power systems. Discussions with Mr. P. Pentayya, General Manager, Power System Operation Corporation (POSOCO) of India, Mr. J.D. Kulkarni, Chief Operating Officer, Tata Power Trading Company and Dr. K. Rajamani, Chief Consultant, Reliance Infrastructures Limited were of great help. We sincerely thank them for clarifying important operational aspects of power systems. We wish to make it clear that commissions and omissions, if any, in formulation or in the text are our own.

Authors were unable to trace the copyright (for permissions) for Report of The Task Force on Technical Study in Regard to Grid Stability (From 23 July to 31 July 2012). A map and some data from this report have been reproduced in the lecture notes for use in illustrative examples. Our thanks are due to appropriate authorities for making this report available on the net.

We hereby acknowledge support from Indian Institute of Technology Bombay (IITB) and Veermata Jijabai Technical Institute (VJTI), Mumbai in this work.

Authors are grateful to members of their families for their encouragement and patience.

We are thankful to M/s. Springer for undertaking publication of this work.

Contents

About the Authors

Prof. M.V. Hariharan is Gold Medal awardee for his performance in B.Tech. in the entire state of Madras (1954) and a UNESCO scholar (1959). He has been a popular faculty at IIT Bombay, Mumbai (1962–1993) and is known for his down-to-earth teaching methods. He has been Visiting Professor at University of Manitoba, Canada (1987) and at University of Western Ontario London, Canada (1988). He was consultant to Tata Consulting Engineers, Maharashtra State Electricity Board, now restructured as Mahagenco (MSPGCL), Mahatransco (MSETCL) and Mahadiscom (MSEDCL). Professor Hariharan has carried out research projects for Department of Science and Technology, India and has published papers in international journals. He has guided many Ph.D. students.

Prof. S.D. Varwandkar Ph.D. (IIT Kanpur, India) was faculty at VJTI, Mumbai (1969–2004) and has taught courses on electrical machines, power system analysis, and planning and reforms and has published in international journals and guided three Ph.D. students. He has carried out projects for Board of Research in Nuclear Sciences, India and was Expert Consultant at Global R&D, Crompton Greaves, Mumbai (2007–2008).

Ms. Pragati P. Gupta topped in BE among all branches of Engineering at Dibrugarh University in 1991 and in M.Tech. (Power Systems) at VJTI. She is currently Assistant Professor of Electrical Engineering at VJTI and is associated with teaching of courses on power systems, especially deregulated systems.

Abbreviations and Symbols

V	Nodal vector of voltages
I	Nodal vector of currents
Y	Bus admittance matrix
Z	Bus impedance matrix
Z_{LN}	Line impedance
Z_{LD}	Load impedance
A	Node-element incidence matrix
$\tilde{A}$	Node-element incidence matrix including ground node
ng	Number of generators
nd	Number of loads
nl	Number of transmission lines
v	Vector of element voltages
V_{ii}	Voltage at node i when only one generator i feeds the network
I_{ii}	Current at node i when only one generator i feeds the network
z_e	Impedance of element e
y_e	Admittance of element e
KS_i	*Kirchhoff State* for generator i
v_{ei}	(*Kvoltage*), magnitude of *voltage across* element e in Kirchhoff State i
i_{ei}	(*Kcurrent*), magnitude of *current* in element e in Kirchhoff state i
p_{ei}	(*Kpower*), $i_{ei}^2 r_e$, of element e in Kirchhoff State i
v_e	Net magnitude of *voltage across* element e
i_e	Net magnitude of *current* in element e
p_e	$i_e^2 r_e$ real power of element e
q_e	$i_e^2 x_e$ reactive power of element e
Λ_i	Module of generator i
E_{ei}	Electrical description, also 'Hilbert space' of element module e in Kirchhoff State i
ε_{ei}	*Loss* power fractions for e–i pair
ε_{eif}	*Flow* power fractions for e–i pair
η_{ei}	Reactive power *loss* fractions for e–i pair

η_{eif}	Reactive power *flow* fractions for *e–i* pair
P_{eif-in}	Power flow *into* the element *e* in Kirchhoff State *i*
$P_{eif-out}$	Power flow *out of* the element *e* in Kirchhoff State *i*
P_{eif}	Power flow *on* the element *e* in Kirchhoff State *i*
P_{ef-in}	Net power flow *into* the element *e*
P_{ef-out}	Net power flow *out of* the element *e*
P_{ef}	Net power flow *on* the element *e*
Γ_{ei}	Orthoframe for pair *e–i*
Γ_{i}	Orthoframe for generator *i*
$P_{fe}^{\max}$	Upper limit of power flow on line *e*
P_{fre}	Flow reserve (margin) available on line *e*
ξ	Xi Matrix
BIS	Blackout incipient
CERC	Central Electricity Regulatory Commission
DSG	Dispersed System Generation
FACTS	Flexible AC Transmission System
HVDC	High Voltage Direct Current
ILF	Iterative load flow
LP	Linear programming
MLF	Modular load flow
RM	Radial mesh
SLFE	Static load flow equations

Chapter 1
Introduction

Topology is the property of something that doesn't change when you bend it or stretch it as long as you don't break anything.
Edward Witten

1.1 Review of Graph Theory

Graph theory is a vast mathematical discipline used in various engineering fields. The first paper on graphs was written by famous Swiss mathematician Euler [1]. Kirchhoff was first to use graph theoretical concepts to characterize electrical network in the paper he published in 1847 [2]. Graph theory has evolved into an important mathematical tool in solution of a wide variety of engineering problems [3]. Power system is a widespread large complex network. It is a collection of physical components and devices interconnected electrically. For large networks, graph theory has been used for easy formulation of equations [4–6]. Only selected topics of this theory applicable to our work are mentioned here.

Graph of a network captures its topology i.e. aspects related to its connectivity. It is a useful representation and generalisation of a network. Many network equations are invariant across networks with the same topology. This includes equations derived from Kirchhoff's laws. The significance of graph theory is illustrated with following example. Figure 1.1 represents two network segments of two different networks and their structure is represented by a graph. It is seen from Fig. 1.1, that KCL at node in network segments (i) and (ii) as well as in (iii) yields $I_1 + I_2 = I_3$. This implies network equations derived from KCL or KVL depend on structure of the network rather than type of elements. For large complex networks, graph theory simplifies formulation of network equations as we need to consider only structure i.e. topology of network.

M.V. Hariharan et al., *Modular Load Flow for Restructured Power Systems*,
Lecture Notes in Electrical Engineering 374, DOI 10.1007/978-981-10-0497-1_1

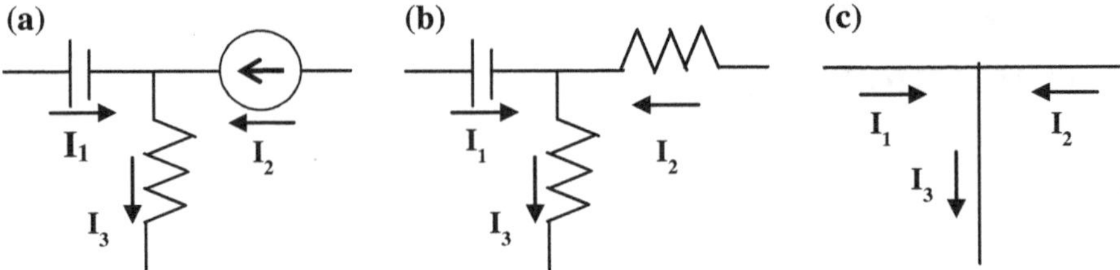

Fig. 1.1 Segment extracted from different networks (**a**) and (**b**)

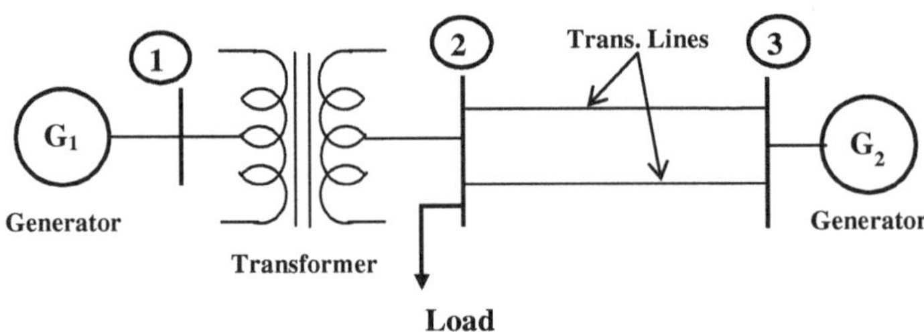

Fig. 1.2 A typical power system network

To describe geometrical feature of a network, it is replaced by single line segments called elements and whose terminals are called nodes. Terms related to the graph theory are discussed with typical power system network (Fig. 1.2).

In graph, element numbers are usually not circled while node numbers are usually, but not necessarily, circled. Graph, in which elements are assigned with arbitrary direction, is called oriented graph. In power networks, out of n nodes, one node which is at zero potential is assigned a number, usually zero.

Rank of a Graph: The rank of graph consisting of n nodes is $(n - 1)$. For graph of Fig. 1.3, rank is 3 (i.e., 4 − 1). The oriented graph is shown in Fig. 1.4.

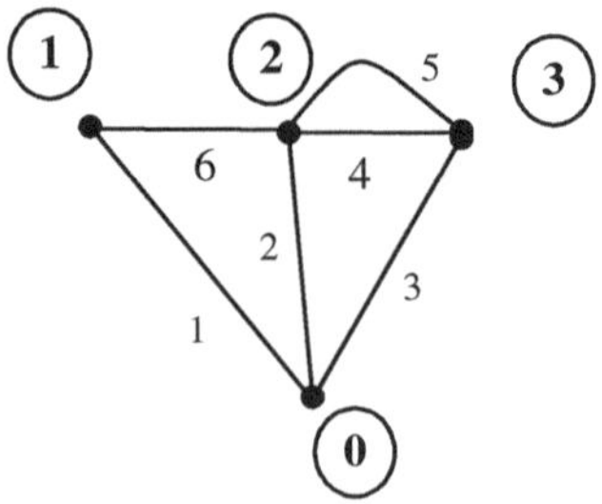

Fig. 1.3 Graph of power system network of Fig. 1.2

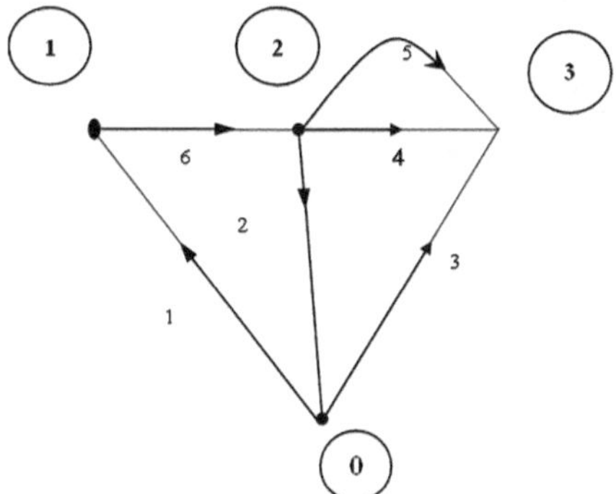

Fig. 1.4 Oriented graph

Sub-graph: It is a subset of the nodes and elements of a graph.

Connected Graph or Sub-graph: A graph or sub-graph is said to be connected when there exists at least one path between any two nodes.

Tree: A connected sub-graph containing all nodes of graph but having no closed paths is called tree. The elements of tree are called *branches or twigs*. The number of branches, $b = n - 1$, where n is total number of nodes (Fig. 1.5).

Co-Tree: A sub-graph of graph which is formed with the elements other than those in the tree. Elements of co-tree are called *links or chords*. Co-tree is not necessarily connected (Fig. 1.6). The number of links l of a connected graph with e elements is

$$l = e - b = e - n + 1 \tag{1.1}$$

Trees and co-trees of a graph are not unique.

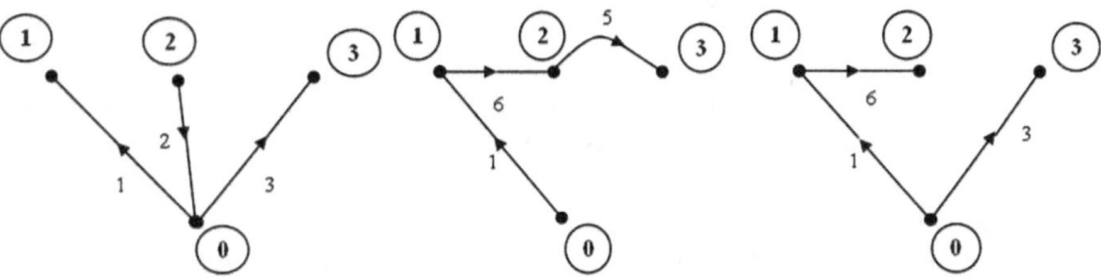

Fig. 1.5 Some possible trees of given graph

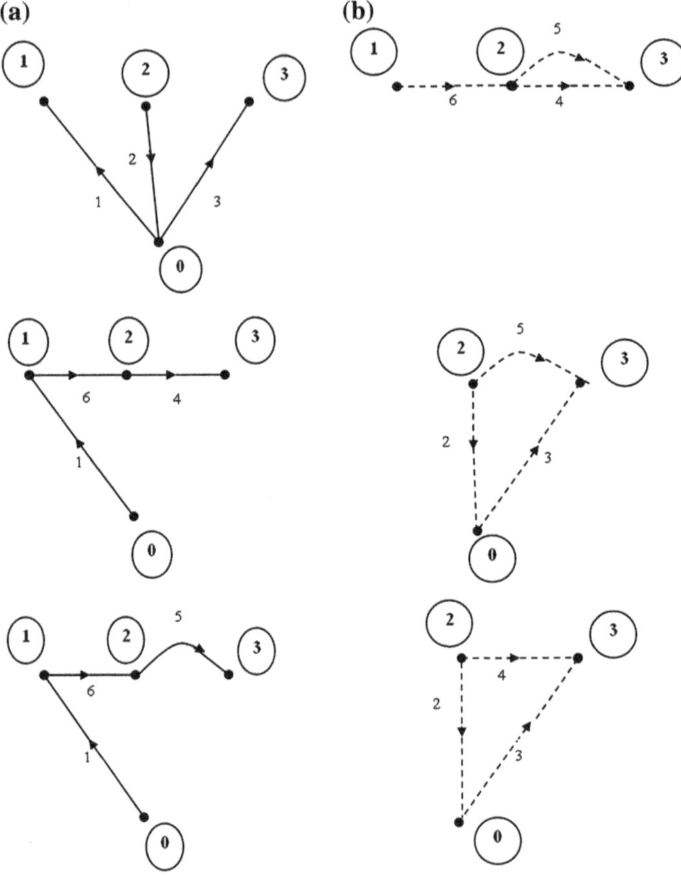

Fig. 1.6 Some possible trees of given graph and corresponding co-tree

1.1.1 Fundamental Cutsets (f-Cutsets)

Fundamental or basic cutset of a graph is set of elements consisting of only *one* branch and a number of links. Removal of these elements from graph results in division of graph into two unconnected sub-graphs. An isolated node is considered as a sub-graph in this process (Fig. 1.7).

Number of fundamental f-cutsets = Number of branches.

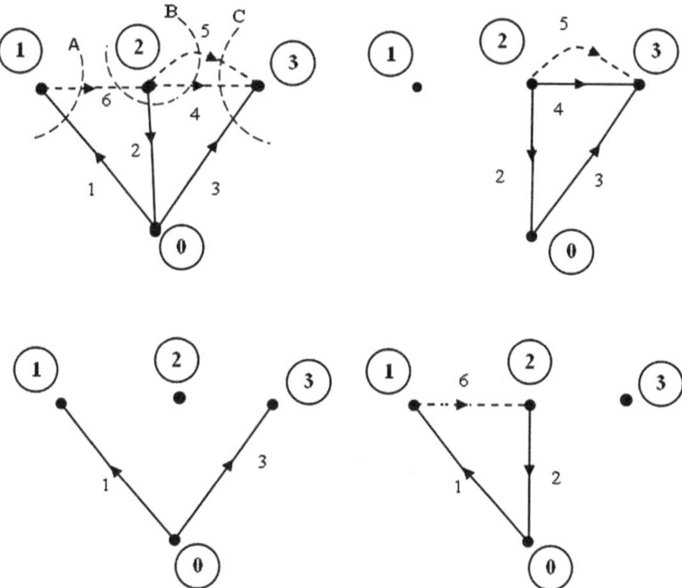

Fig. 1.7 Basic or f-cutsets of the graph

1.1.2 Incidence Matrices

Each element of a graph is incident to two nodes. Incidence matrices provide information about the incidence of elements to nodes and their orientation with respect to nodes. There are various incidence matrices defined to give information of incidence of elements to nodes (buses), loops, cutsets and paths of a graph. Node-element incidence matrix and the bus incidence matrix are defined below.

Node element incidence matrix ($\tilde{A}$, *and*, A): This matrix gives information whether element is incident to a particular node or not. In a column it has two non-zero entries corresponding to two nodes to which an element is incident.

Elements a_{ij} of matrix of $\tilde{A}$ or A are found as per following rules:

$a_{ij} = 1$, if the ith element is incident to and oriented away from the jth node
$= -1$, if the ith element is incident to and oriented towards jth node
$= 0$, if the ith element is not incident to the jth node

For a given power system network (Fig. 1.2) and its oriented graph (Fig. 1.4), element node incidence matrix $\tilde{A}$ includes row corresponding to the ground node.

$$\tilde{A} = \begin{array}{c} \textit{elements} \rightarrow \\ \\ \textit{nodes} \downarrow \\ \\ \end{array} \begin{array}{c} \\ 1 \\ 2 \\ 3 \\ 0 \end{array} \begin{array}{c} \begin{matrix} 1 & 2 & 3 & 4 & 5 & 6 \end{matrix} \\ \begin{bmatrix} -1 & 0 & 0 & 0 & 0 & 1 \\ 0 & 1 & 0 & 1 & 1 & -1 \\ 0 & 0 & -1 & -1 & -1 & 0 \\ 1 & -1 & 1 & 0 & 0 & 0 \end{bmatrix} \end{array} \tag{1.2}$$

Elimination of row corresponding to the ground node is called *bus incidence matrix* in power systems. However we will call this also *node-element incidence* matrix but denote it by A (without *tilde*).

$$A = \textit{nodes} \downarrow \begin{array}{c} \\ 1 \\ 2 \\ 3 \end{array} \begin{array}{c} \textit{elements} \rightarrow \begin{matrix} 1 & 2 & 3 & 4 & 5 & 6 \end{matrix} \\ \begin{bmatrix} -1 & 0 & 0 & 0 & 0 & 1 \\ 0 & 1 & 0 & 1 & 1 & -1 \\ 0 & 0 & -1 & -1 & -1 & 0 \end{bmatrix} \end{array} \tag{1.3}$$

1.1.3 Kirchhoff's Laws

KCL for currents at nodes:
$\tilde{A}\, i = 0$; i is vector of element currents
KVL for element voltages:
$v = \tilde{A}^T V$; v is vector of element voltages, and, V is vector of node voltages.

1.1.4 Primitive Network

A power network consists of number of elements which are incident between any two nodes. These elements may be a purely active (voltage source or current source) or purely passive (admittance or impedance) or combination of both. Each element can be represented by decoupled partial network, either in impedance or admittance form. The decoupled partial network is known as primitive network.

The impedance form is a voltage source E_{jk} in series with an impedance z_{jk}. Admittance form is represented with current source J_{jk} in parallel with an admittance y_{jk}. The element current is i_{jk} and element voltage is $v_{jk} = E_j - E_k$; E_j and E_k are voltages at element nodes j and k respectively.

The performance equation for primitive network in impedance form from Fig. 1.8a is

$$E_{jk} + v_{jk} = z_{jk} i_{jk} \tag{1.4}$$

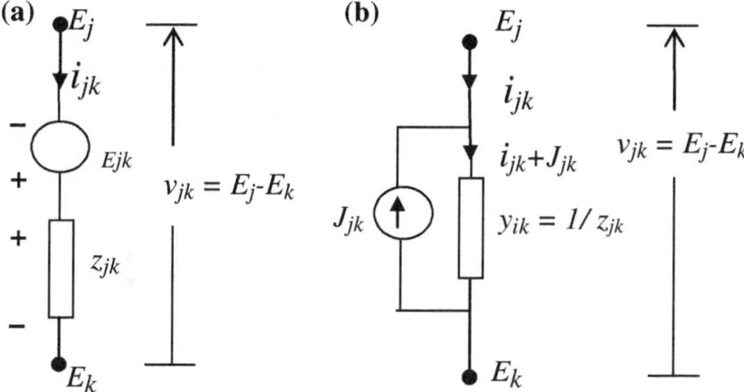

Fig. 1.8 Primitive network in impedance form (**a**) and admittance form (**b**)

Similarly for admittance form from Fig. 1.8b, performance equation is

$$i_{jk} + J_{jk} = y_{jk}v_{jk} \tag{1.5}$$

For all elements vector performance equation in impedance form can be written as

$$v + E = zi \tag{1.6}$$

In admittance form,

$$i + J = yv \tag{1.7}$$

where, v and i are element voltage and current vectors respectively. E and J are voltage and current *source* vectors respectively. z is primitive impedance matrix; y is primitive admittance matrix; z and y are diagonal matrices. The diagonal entries are element impedance/admittances.

1.1.5 Network Structure

Node element incidence matrix A describes the structure of a network. It has rows equal to number of buses and columns equal to number of elements in a network. For network shown in Fig. 1.9 and its oriented graph in Fig. 1.10, we convert load at bus 3 into impedance and include it in the network. The A matrix for this network given in (1.8) will be used in Chap. 4

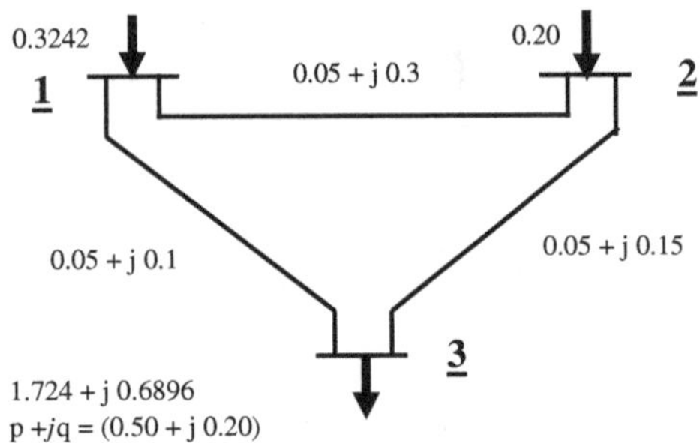

Fig. 1.9 A 3-bus network with impedance values

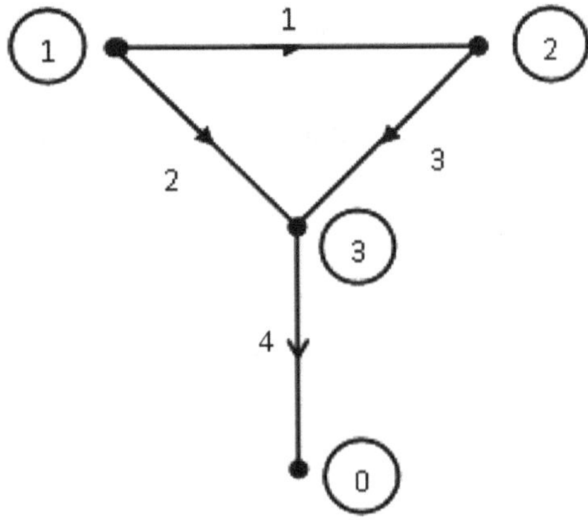

Fig. 1.10 Oriented graph of bus network of Fig. 1.9

$$
\begin{array}{c} \\ \\ \end{array}
\begin{array}{cccc} Element\,1 & Element\,2 & Element\,3 & Element\,4 \\ Line\,1-2 & Line\,1-3 & Line\,2-3 & Load\,@3 \end{array}
$$

$$
A = \begin{array}{c} Bus\,1 \\ Bus\,2 \\ Bus\,3 \end{array}\begin{bmatrix} 1 & 1 & 0 & 0 \\ -1 & 0 & 1 & 0 \\ 0 & -1 & -1 & 1 \end{bmatrix} \tag{1.8}
$$

1.1.6 Network Matrices

We illustrate these by examples. *Primitive admittance matrix* for the network shown in Fig. 1.9 is given by,

$$y = \begin{bmatrix} 0.5405 - j3.2432 & 0 & 0 & 0 \\ 0 & 4.000 - j8.000 & 0 & 0 \\ 0 & 0 & 2.000 - j6.000 & 0 \\ 0 & 0 & 0 & 0.50 - j0.20 \end{bmatrix} \tag{1.9}$$

Bus admittance matrix is computed from the expression,

$$Y = AyA^T \tag{1.10}$$

The network equations in admittance form are written as,

$$I = YV \tag{1.11}$$

Corresponding impedance form is,

$$V = ZI; \quad Z = Y^{-1} \tag{1.12}$$

Vector I is nodal current injection vector. Matrix Z is called the bus impedance matrix. We display below the Z matrix for the network shown in Fig. 1.9. Load is considered as an element of the network.

$$Z = \begin{bmatrix} 1.7607 + j0.7723 & 1.7395 + j0.7166 & 1.7241 + j0.6897 \\ 1.7395 + j0.7166 & 1.7580 + j0.7989 & 1.7241 + j0.6897 \\ 1.7241 + j0.6897 & 1.7241 + j0.6897 & 1.7241 + j0.6897 \end{bmatrix} \tag{1.13}$$

1.1.7 Power Network

Electric circuit of a power network has special features which distinguish it from general electric circuits. Its transmission line elements transfer power from generators to loads which are special circuits that consume most of the generator powers. The concept of power transfer is not natural to general circuits where consumption or loss is the relevant entity: there is nothing like *power-flows*. Loads in power systems are special circuits which consist of motors, devices, apparatus and electrical appliances that consume electricity and are designed for standard voltages. Transmission lines are designed for high efficiency and low voltage regulation which means that the line impedances are very small as compared to those of loads. Neglecting transmission line resistances therefore does not affect overall results significantly. Maintaining slightly higher than the nominal voltage at sources would result in close-to-nominal voltage at load buses under normal loading conditions.

1.2 Iterative Load Flow

Static load flow equations (SLFE) [7] in iterative load flow (ILF) are written in terms of bus voltages and their phase angles. One of the buses is called swing, reference, or, slack bus and is assigned a fixed angle (usually 0). Objective of ILF is to obtain voltages and their phase angles that satisfy these equations. These being nonlinear, their solution requires numerical iterative techniques. Modular Load Flow (hereafter we shorten it as MLF) proposed in the book does not work with SLFE. It is non-iterative i.e., the solution for flows, losses and voltages is obtained from explicit expression.

1.3 Issues with Iterative Load Flow

1.3.1 Premise

A statement is usually made while formulating ILF that specifying two variables at each bus ensures a solution since we get two independent equations in two unknowns at each bus. The statement is true only for *linear* equations. The jacobian can indeed become singular at times; indicating that the SLFEs do become dependent equations during iterations and computation thereafter is not possible. In reality however, voltages and currents, whatever their magnitudes, must appear in a passive circuit if sources are applied to it. MLF provides this solution.

1.3.2 Voltage Specification at PV Buses

In ILF it may be necessary to specify voltages which cannot be physically maintained constant due to reactive power shortage. This is one of the reasons for its non-convergence. Also, for every voltage specification (without any reality check) the analyst will get a different solution. Load flow solution is thus a *possibility* and not a real physical solution.

Homotopic and Holomorphic methods [8, 9] try to override this problem in different ways. MLF is still another formulation and obviates apriori specification of voltages.

1.4 Load Representation

Loads are represented by constant P and Q in ILF. These are represented by constant impedances in MLF. Both representations are approximations. One great advantage of constant impedance representation is that the entire power network becomes a linear circuit and admits circuit solution procedures with power acting as source. We explain this in Chap. 2.

1.5 Concept of Port

Concept of port is not new to networks in which elements are interconnected to form nodes. In graph theory, nodes are basic entities and admit an independent description. A *port* on the other hand is basic to network theory. The most elementary port consists of a single R-X (in series) circuit element. It may receive power from a single source or from multiple sources. We can associate three quantities with a port; real power, voltage across the element, and the current through it. These three variables uniquely determine role of element in the circuit. With excitations at multiple ports, individual source contributions can be superposed *appropriately* to obtain net voltages, currents, and powers in the element. It is desirable for readers to become comfortable with the use of port (element) variables for analysis in order to understand the Modular Load Flow.

1.6 Matrix ξ

We introduce a new construct called the ξ(Xi) matrix derived from the conventional bus impedance matrix Z. Columns of ξ are of type $Z_{mi} - Z_{ni}$ where i is column index of generators and m and n are end nodes of element e. For *element* 1:1–2; *element* 2:1–3; *element* 3:2–3 and *element* 4:3–0, in Fig. 1.9, the ξ matrix is given by,

$$\xi = \begin{array}{r} element\,1 \\ element\,2 \\ element\,3 \\ element\,4 \end{array} \left[\begin{array}{c|c} Z_{11} - Z_{21} & Z_{12} - Z_{22} \\ \hline Z_{11} - Z_{31} & Z_{12} - Z_{32} \\ \hline Z_{21} - Z_{31} & Z_{22} - Z_{32} \\ \hline Z_{33} & Z_{33} \end{array} \right] \tag{1.14}$$

Expressed in shorter notation,

$$\xi = \left[\begin{array}{c|c} \xi_{11} & \xi_{12} \\ \hline \xi_{21} & \xi_{22} \\ \hline \xi_{31} & \xi_{32} \\ \hline \xi_{41} & \xi_{42} \end{array} \right] \tag{1.15}$$

Matrix ξ provides decoupled representation with respect to generators on one hand and the element voltages on the other. With only one source applied at a time, the element voltages are,

$$v_1 = \xi_1 I_{11} = \begin{bmatrix} \xi_{11} \\ \xi_{21} \\ \xi_{31} \\ \xi_{41} \end{bmatrix} I_{11}, \quad \text{and,} \quad v_2 = \xi_2 I_{22} = \begin{bmatrix} \xi_{12} \\ \xi_{22} \\ \xi_{32} \\ \xi_{42} \end{bmatrix} I_{22}, \text{etc.} \tag{1.16}$$

i.e.,

$$v_i = \xi_i I_{ii}$$
$$v_i = vector\ of\ element\ voltages\ corresponding\ to\ generator\ i$$
$$I_{ii} = current\ injected\ by\ generator\ i,\ when\ acting\ alone$$

It is easy to see that multiplier column is given by,

$$\xi_i = \left[A^T Z\right]_{col,i} \tag{1.17}$$

Voltage components in a single element due to individual sources will have two subscripts. For example, voltage across element e due to injected current at generator bus i (all others set to zero) is,

$$v_{ei} = \xi_{ei} I_{ii} \tag{1.18}$$

Matrix ξ will play a crucial role in development of the Modular Load Flow.

References

1. L. Euler, Solutio problematis ad geometriam situs pertinentis. Novi Commentarii Acad. Scientarium Imperialis Petropolitanque **7**(1758–1759), 9–28.
2. G. Kirchhoff, Ueber die Auflösung der Gleichungen, auf welche man bei der Untersuchung der linearen Vertheilung galvanischer Ströme geführt wird. Ann. Phys. **148**, 497–508 (1847). doi:10.1002/andp.18471481202
3. K.V.V. Murthy, M.S. Kamath, *Basic Circuit Analysis*, 8th edn. (Jaico Publishing House India, 2010)
4. G.W. Stagg, A.H. El-Abiad, *Computer Methods in Power System Analysis*, 1st edn. (McGraw-Hill Inc., US, 1968)
5. M.A. Pai, *Computer Techniques in Power System Analysis* (Tata McGraw-Hill, New Delhi, 1979)
6. G.T. Heydt, *Computer Analysis Methods for Power Systems*, 2 Edn. (Stars in a Circle Publication, 1994)
7. O.I. Elgerd, *Electric Energy Systems Theory* (Tata-McGraw-Hill, Delhi, 1983)
8. D. Cai, Y. Chen, Application of Homotopic methods to power systems. J. Comput. Math. **22**(1), 61–68
9. A. Trias, The Holomorphic embedding load flow method. *IEEE PES General Meeting*, July 2012

Chapter 2
Circuit Solutions

Old order changeth yielding place to new. And God fulfils himself in many ways lest one good custom should corrupt the world...

Alfred Tennyson

In this chapter we discuss solution of circuits with current and voltage sources and develop a new procedure which uses power source. Voltage/current sources are well-known in literature but *power* as source does not find mention in any of the books on electric circuits (e.g., see [1]). *Power source* however is a reality in power networks. To be able to do this, variables in our analysis will be *element* voltages and currents and not *nodal* voltages and currents. Since element variables have not been customarily used in the past for general power system analysis, it will be necessary to remind ourselves of this fact occasionally, if not frequently.

2.1 Multi-terminal Representation

Consider a system with nb buses. The conventional bus voltage and current vectors are related by the equation [2],

$$V = \begin{bmatrix} Z_{11} & Z_{12} & & Z_{1i} & & Z_{1,nb} \\ Z_{21} & Z_{22} & & Z_{2i} & & Z_{2,nb} \\ . & . & \cdots & . & \cdots & . \\ . & . & & . & & . \\ Z_{nb,1} & Z_{nb,2} & & Z_{nb,i} & & Z_{nb,nb} \end{bmatrix} \begin{bmatrix} I_1 \\ I_2 \\ . \\ . \\ I_i \\ . \\ I_{nb} \end{bmatrix} \tag{2.1}$$

M.V. Hariharan et al., *Modular Load Flow for Restructured Power Systems*, Lecture Notes in Electrical Engineering 374, DOI 10.1007/978-981-10-0497-1_2

$$= \begin{bmatrix} Z_{11} \\ Z_{21} \\ \cdot \\ \cdot \\ Z_{nb,1} \end{bmatrix} I_1 + \begin{bmatrix} Z_{12} \\ Z_{22} \\ \cdot \\ \cdot \\ Z_{nb,2} \end{bmatrix} I_2 + \cdots + \begin{bmatrix} Z_{1,nb} \\ Z_{2,nb} \\ \cdot \\ \cdot \\ Z_{nb,nb} \end{bmatrix} I_{nb} \tag{2.2}$$

We can write (2.2) as,

$$V = V(1) + V(2) + \cdots + V(nb) \tag{2.3}$$

where,

$$V(k) = \begin{bmatrix} V_{1k} \\ \hline V_{2k} \\ \hline \cdot \\ \hline \cdot \\ \hline V_{nb,k} \end{bmatrix} \tag{2.4}$$

In what follows, we will use Z-bus and consider a single source. This implies using only one column of the bus impedance matrix Z. This idea is different from that in basic circuit analysis that employs node analysis or mesh analysis to obtain circuit currents and node voltages [1]. For multiple sources superposition poses an interesting challenge.

2.2 Solution for Element Variables

2.2.1 Single Current Source

Consider only *one* source current injected at bus *i*. Set all other injections to zero. Let I_{ii} be the source current. (We will reserve single subscripted I_i and V_i for current and voltage when multiple sources are connected.) Voltages of buses *m* and *n* are given by,

$$V_{mi} = Z_{mi} I_{ii} \tag{2.5}$$

$$V_{ni} = Z_{ni} I_{ii} \tag{2.6}$$

Voltage *across* and the current *in* an element *e* (between buses *m* and *n*) are,

$$\begin{aligned} v_{ei} &= V_{mi} - V_{ni} = (Z_{mi} - Z_{ni}) I_{ii} \\ v_{ei} &= \xi_{ei} I_{ii} \end{aligned} \tag{2.7}$$

$$i_{ei} = y_e (V_{mi} - V_{ni}) = y_e \xi_{ei} I_{ii} \tag{2.8}$$

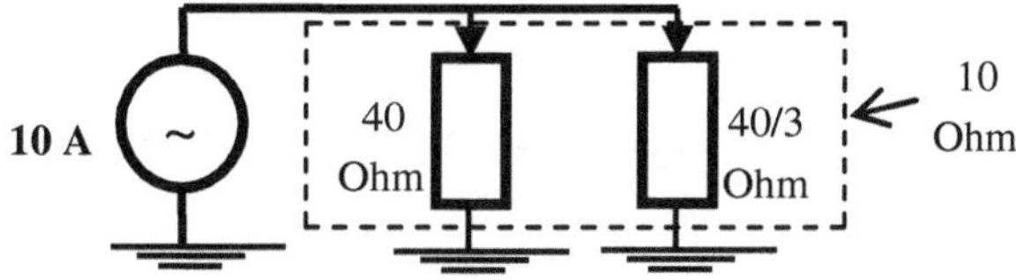

Fig. 2.1 Circuit with current source

Term ξ_{ei} has been defined in (1.17).
Power in element e due to generator i is,

$$\begin{aligned} p_{ei} &= \mathrm{Re}\{v_{ei}i_{ei}^*\} \\ &= |I_{ii}|^2\mathrm{Re}\{\xi_{ei}\xi_{ei}^* y_e^*\} \end{aligned} \tag{2.9}$$

2.2.2 Example

Consider the circuit in Fig. 2.1. Net resistance offered to the source is,

$$40 \| \frac{40}{3} = 10\,\Omega$$

$$\begin{aligned} \textit{Source Current} &= 10\,\mathrm{A} \\ \textit{Voltage across loads} &= \textit{common bus voltage} = 10 \times 10 = 100\,\mathrm{V} \end{aligned}$$

Resistance values: *element* 1 $= 40\,\Omega$, *element* 2 $= \frac{40}{3}\,\Omega$
Bus impedance and node element matrices are,

$$Z = [10] : A = [1 \quad 1]$$

From (1.15) and (1.17),

$$\xi \underline{\underline{\Delta}} \begin{bmatrix} \xi_{11} \\ \xi_{21} \end{bmatrix} = A^T Z = \begin{bmatrix} 1 \\ 1 \end{bmatrix} [10] = \begin{bmatrix} 10 \\ 10 \end{bmatrix}$$

Using (2.7) and (2.8),

$$\begin{aligned} \begin{bmatrix} v_{11} \\ v_{21} \end{bmatrix} &= \begin{bmatrix} \xi_{11} \\ \xi_{21} \end{bmatrix} I_i = \begin{bmatrix} 10 \\ 10 \end{bmatrix} 10 = \begin{bmatrix} 100 \\ 100 \end{bmatrix} \mathrm{V} \\ \begin{bmatrix} i_{11} \\ i_{21} \end{bmatrix} &= \begin{bmatrix} y_1 & 0 \\ 0 & y_2 \end{bmatrix} \begin{bmatrix} v_{11} \\ v_{21} \end{bmatrix} = \begin{bmatrix} \frac{1}{40} & 0 \\ 0 & \frac{3}{40} \end{bmatrix} \begin{bmatrix} 100 \\ 100 \end{bmatrix} \\ &= \begin{bmatrix} 2.5 \\ 7.5 \end{bmatrix} \mathrm{A} \end{aligned}$$

Using (2.9),

$$Power\, in\, 40\,\Omega = (10)^2 \times (10 \times 10) \times \frac{1}{40} = 250\,\mathrm{W}$$

Similarly,

$$Power\, in\, \frac{40}{3}\,\Omega = (10)^2 \times (10 \times 10)\frac{3}{40} = 750\,\mathrm{W}$$
$$Total\, Power = 250 + 750 = 1000\,\mathrm{W}$$

From total current and net resistance,

$$Total\, Power = 10^2 \times 10 = 1000\,\mathrm{W}$$

2.2.3 Single Voltage Source

In this case,

$$|I_{ii}| = \left|\frac{V_{ii}}{Z_{ii}}\right| \tag{2.10}$$

$$v_{ei} = V_{mi} - V_{ni} = (Z_{mi} - Z_{ni})I_{ii} = \xi_{ei} I_{ii} \tag{2.11}$$

$$i_{ei} = y_e(V_{mi} - V_{ni}) = y_e \xi_{ei} I_{ii} \tag{2.12}$$

And element power expression (2.9) becomes,

$$p_{ei} = \left|\frac{V_{ii}}{Z_{ii}}\right|^2 \mathrm{Re}\{\xi_{ei}\xi_{ei}^* y_e^*\} \tag{2.13}$$

2.2.4 Example

Consider the circuit in Fig. 2.2. Let the voltage applied be 100 V.

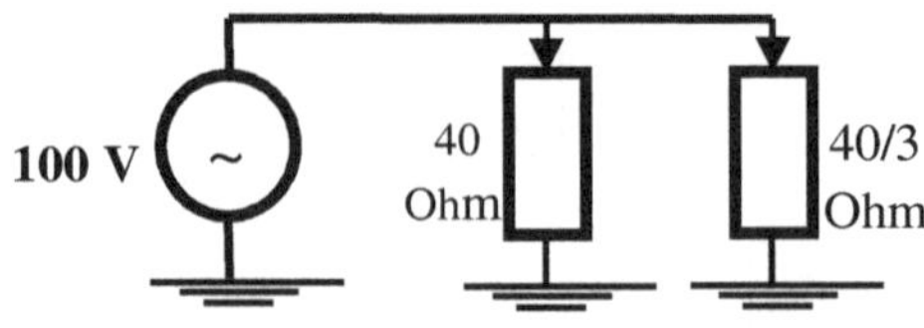

Fig. 2.2 Circuit with voltage source

Net resistance is, $40 \| \frac{40}{3} = 10\,\Omega$

$$\begin{aligned}
Source\,Current &= \frac{100}{10} = 10\,\text{A} \\
Z &= [10]; A = [1 \quad 1] \\
\xi &= \begin{bmatrix} \xi_{11} \\ \xi_{21} \end{bmatrix} = A^T Z = \begin{bmatrix} 1 \\ 1 \end{bmatrix} [10] = \begin{bmatrix} 10 \\ 10 \end{bmatrix} \\
\begin{bmatrix} v_{11} \\ v_{21} \end{bmatrix} &= \begin{bmatrix} \xi_{11} \\ \xi_{21} \end{bmatrix} I_i = \begin{bmatrix} 10 \\ 10 \end{bmatrix} 10 = \begin{bmatrix} 100 \\ 100 \end{bmatrix} \text{V} \\
\begin{bmatrix} i_{11} \\ i_{21} \end{bmatrix} &= \begin{bmatrix} y_1 & 0 \\ 0 & y_2 \end{bmatrix} \begin{bmatrix} v_{11} \\ v_{21} \end{bmatrix} = \begin{bmatrix} \frac{1}{40} & 0 \\ 0 & \frac{3}{40} \end{bmatrix} \begin{bmatrix} 100 \\ 100 \end{bmatrix} \\
&= \begin{bmatrix} 2.5 \\ 7.5 \end{bmatrix} \text{A}
\end{aligned}$$

Using (2.13),

$$Power\,in\,40\,\Omega = \left(\frac{100}{10}\right)^2 \times (10 \times 10) \times \frac{1}{40} = 250\,\text{W}$$

Similarly,

$$\begin{aligned}
Power\,in\,\frac{40}{3}\,\Omega &= \left(\frac{100}{10}\right)^2 \times (10 \times 10) \frac{3}{40} = 750\,\text{W} \\
Total\,Power &= 250 + 750 = 1000\,\text{W}
\end{aligned}$$

From total current and net resistance,

$$Total\ Power = 10^2 \times 10 = 1000\,\text{W}$$

2.2.5 Single Power Source

In this case,

$$I_{ii}^2 = \frac{P_i}{R_{ii}}; R_{ii} = \text{Re}\{Z_{ii}\} \tag{2.14}$$

Therefore,

$$I_{ii} = \sqrt{\frac{P_i}{R_{ii}}} \tag{2.15}$$

$$v_{ei} = V_{mi} - V_{ni} = \xi_{ei} I_{ii} \tag{2.16}$$

$$i_{ei} = y_e(V_{mi} - V_{ni}) = y_e \xi_{ei} I_{ii} \tag{2.17}$$

And power in element e due to generator i,

$$p_{ei} = \frac{P_i}{R_{ii}} \mathrm{Re}\{\xi_{ei}\xi_{ei}^* y_e^*\} \tag{2.18}$$

2.2.6 Example

Consider power source as shown in Fig. 2.3. Let the power injected be 1000 W. For it to be expended in the net 10 Ω, the current will be,

$$\begin{aligned}
Source\ Current &= \sqrt{\frac{1000}{10}} = 10\text{A} \\
Z &= [10]; A = [1 \quad 1] \\
\xi &= \begin{bmatrix} \xi_{11} \\ \xi_{21} \end{bmatrix} = A^T Z = \begin{bmatrix} 1 \\ 1 \end{bmatrix} [10] = \begin{bmatrix} 10 \\ 10 \end{bmatrix} \\
\begin{bmatrix} v_{11} \\ v_{21} \end{bmatrix} &= \begin{bmatrix} \xi_{11} \\ \xi_{21} \end{bmatrix} I_i = \begin{bmatrix} 10 \\ 10 \end{bmatrix} 10 = \begin{bmatrix} 100 \\ 100 \end{bmatrix} \text{V} \\
\begin{bmatrix} i_{11} \\ i_{21} \end{bmatrix} &= \begin{bmatrix} y_1 & 0 \\ 0 & y_2 \end{bmatrix} \begin{bmatrix} v_{11} \\ v_{21} \end{bmatrix} = \begin{bmatrix} \frac{1}{40} & 0 \\ 0 & \frac{3}{40} \end{bmatrix} \begin{bmatrix} 100 \\ 100 \end{bmatrix} \\
&= \begin{bmatrix} 2.5 \\ 7.5 \end{bmatrix} \text{A}
\end{aligned}$$

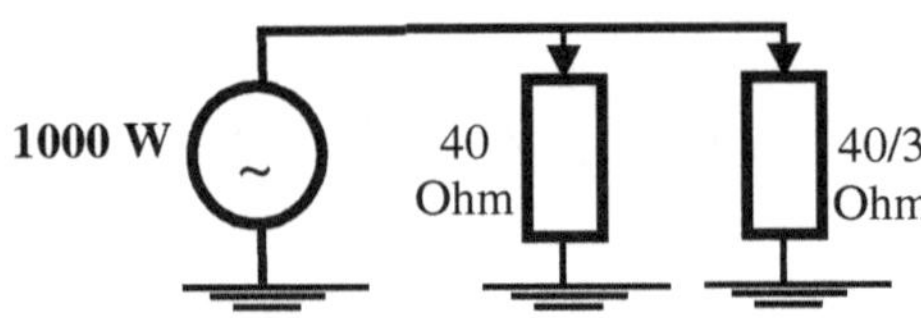

Fig. 2.3 Circuit with power source

Using (2.18),

$$Power\,in\,40\,\Omega = \left(\frac{1000}{10}\right) \times (10 \times 10) \times \frac{1}{40} = 250\,\mathrm{W}$$

Similarly,

$$Power\,in\,\frac{40}{3}\,\Omega = \left(\frac{1000}{10}\right) \times (10 \times 10)\frac{3}{40} = 750\,\mathrm{W}$$
$$Total\,Power = 250 + 750 = 1000\,\mathrm{W}$$

From total current and net resistance, we cross-check that,

$$Total\,Power = 10^2 \times 10 = 1000\,\mathrm{W}$$

Remark Suppose in the above example, the source capacity was limited to 900 W (such limits do arise in ac power systems, e.g., capability curves of turbine generators), then obviously the net current would be $\sqrt{900/10}$ = 9.4868 A, and the voltage, 94.868 V. This depressed voltage (as compared to 100 V with 1000 W) occurs when source is pegged at 900 W. Currents in 40 and 40/3 Ω will be 2.371 A and 7.115 A (for 900 W), as against 2.5 A and 7.5 A (for 1000 W), respectively. Such depressed voltage/current conditions before blackouts often result because of generators reaching thresholds of their *PQ* capabilities. (Capability charts are provided by manufacturers.)

Reflections

Standard solutions with voltage and current sources for passive networks are available in all books on electric circuits. In power network, that we describe here, the source—the synchronous generator—is a 'power' source! No textbook ever mentions constant power as source for electric circuits. We show that solution with 'constant power' source can indeed be obtained and wonder why this has been not been attempted so far..!

References

1. K.V.V. Murthy, M.S. Kamath, *Basic Circuit Analysis*, 8th edn. (Jaico Publishing House India, 2010)
2. M.A. Pai, *Computer Techniques in Power System Analysis* (Tata McGraw-Hill, New Delhi, 1979)

Chapter 3
Kirchhoff State

> *... fundamental concept of node in graph theory is not really as fundamental in network theory and in fact every occurrence of the word 'node' can be replaced by 'port'.*
>
> Prof H. Narayanan [1]

In this chapter a new technique is proposed for solving some special cases of power networks. In practical power systems injected active *powers* act as constant-power sources, because of the fact that they are independently controlled by turbine governors. Here lies the contrast with the conventional stereotype of constant voltage/current type sources. It is well-known that networks, with more than one source, are best solved by superposition, i.e., by solving them with one source at a time and combining individual results to obtain the final one for the original multi-source system. Generators are independent in themselves and hence it stands to reason to think of a subsystem of the power network which is excited by one generator at a time. The subsystem is only an analytical tool. However it constitutes a foundation of our method for solving the special cases mentioned in the beginning. We introduce some new terms to deal with this decoupled view of the system.

The subsystem of a single constant-power source and a number of passive elements will be referred to as Kirchhoff subsystem or a *Ksystem*. Element variables such as voltages, currents of a Ksystem are collectively referred to as Kirchhoff-state variables, or *Kstate-variables*; the voltages being christened as *Kvoltages* and currents as *Kcurrents*. Powers in elements in this subsystem will be called Kirchhoff powers, the *Kpowers*. The state is referred to as a Kirchhoff state, or *Kstate*. *Kstate* is different from the conventional state that is defined by node voltage phasors, with all generators in place. In some sense *Kstate* of a generator is one of the many components of the conventional state. There are as many Kirchhoff states as there are power sources. How the conventional state variables related to the *Kstate-variables* is also discussed in this chapter. Power values are scalars and directly summable over all Ksystems. Simple impedance-voltage triangle relation thereafter gives the net voltages/currents. The orthogonality of electrical variables, discovered in this process is insightful and will be further discussed in Chap. 10.

M.V. Hariharan et al., *Modular Load Flow for Restructured Power Systems*,
Lecture Notes in Electrical Engineering 374, DOI 10.1007/978-981-10-0497-1_3

Our analysis follows from definition of port [1]. A pair of two connected adjacent nodes will be viewed as a port. Every circuit element is thus a port. This concept enables us to formulate a multiport representation for the power network. A generator forms a special port with ground (neutral) and the terminal node constituting the two adjacent nodes. We will use terms *port* and *element* synonymously for power networks.

3.1 Kstate-Variables

In *Ksystem*, element voltages and currents are calculated using circuit solution with power source. Using (2.15), the *Kcurrent* value for injected power P_i is given by,

$$I_{ii} = \sqrt{\frac{P_i}{R_{ii}}} \tag{3.1}$$

R_{ii} is real part of the *i*th diagonal element Z_{ii} of the bus impedance matrix Z. Term Z_{ii} represents the network impedance seen by the *i*th generator.

3.2 Kpowers

From (2.18) the power in element e in Kstate i is,

$$p_{ei} = \frac{P_i}{R_{ii}} \mathrm{Re}\{\xi_{ei}\xi_{ei}^{*}y_e^{*}\} \tag{3.2}$$

$$\xi_{ei} = \left[A^T Z\right]_{\substack{row,\, e \\ col,\, i}} \tag{3.3}$$

Total power loss in an element is the sum of all generator contributions (3.2). There is a little care needed in this summation. In elements of complex network, Kcurrents due to different generators can flow in either direction and could oppose each other depending on the its location vis-à-vis that of the generator. The Kpowers corresponding to opposing flows therefore are *subtracted* while summing. To incorporate this mathematically, we write total power in an element with *sup* before the sigma sign, i.e.,

$$p_e = sup \sum_i p_{ei} \tag{3.4}$$

Flow and loss are considered positive when power flows from the *start* bus m to the *end* bus n of the element. It is considered negative when it flows from n to m.

3.3 Kvoltages

Consider voltage triangle for element e, shown in Fig. 3.1.

With one *power source* the Kvoltage of element e is,

$$v_{ei} = \sqrt{p_{ei}r_e + q_{ei}x_e} = \sqrt{p_{ei}r_e\left(1 + \frac{x_e^2}{r_e^2}\right)} \tag{3.5}$$

And the corresponding Kcurrent is,

$$i_{ei} = y_e v_{ei} \tag{3.6}$$

We illustrate computation of Kstate-variables, Kpowers and the Kstates in Sect. 3.3.1. Kstate-variables will be used to define *modules* in the next chapter.

3.3.1 Example

Consider two resistive loads supplied by two generators as shown in Fig. 3.2.

With only generator 1 (100 W):

Using (3.1),

$$\begin{aligned}
I_{11} &= \sqrt{\frac{100}{10}} = \sqrt{10} \\
V_{11} &= I_{11} * (40\|40/3) = 10\sqrt{10} \\
v_{40,1} &= v_{(40/3),1} = V_{11} = 10\sqrt{10} \\
i_{40,1} &= y_{40}v_{e1} = \frac{1}{40} * (10\sqrt{10}) = \underline{0.7906}\,\text{A} \; ; \\
p_{40,1} &= i_{40,1}^2(40) = 25\,\text{W} \\
i_{40/3,1} &= y_{40/3}v_{(40/3),1} = \frac{3}{40} * (10\sqrt{10}) = \underline{2.3717}\ \text{A}; \\
p_{(40/3),1} &= i_{(40/3),1}^2(40/3) = 75\,\text{W} \\
\textit{Total current with source 1 alone} &= i_{(40,1} + i_{(40/3),1} = 3.1623\,\text{A}
\end{aligned}$$

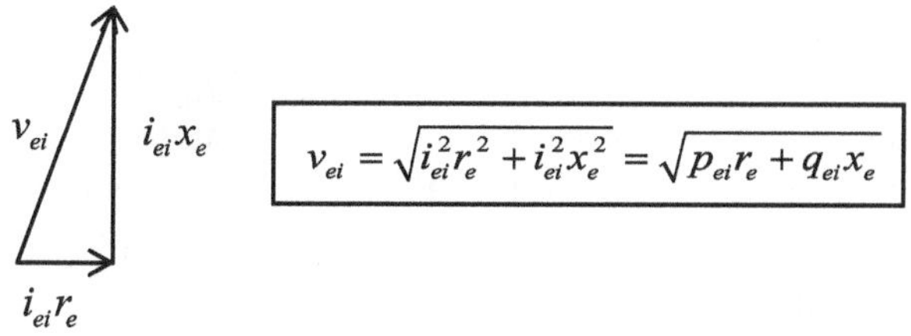

Fig. 3.1 Element voltage

$$v_{ei} = \sqrt{i_{ei}^2 r_e^2 + i_{ei}^2 x_e^2} = \sqrt{p_{ei}r_e + q_{ei}x_e}$$

v_{ei} $i_{ei}x_e$ $i_{ei}r_e$

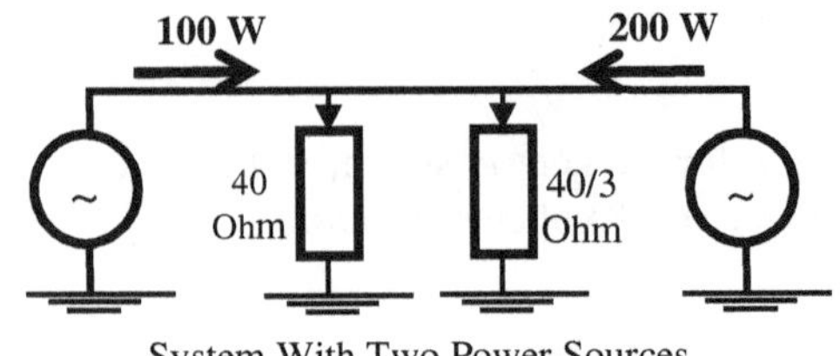

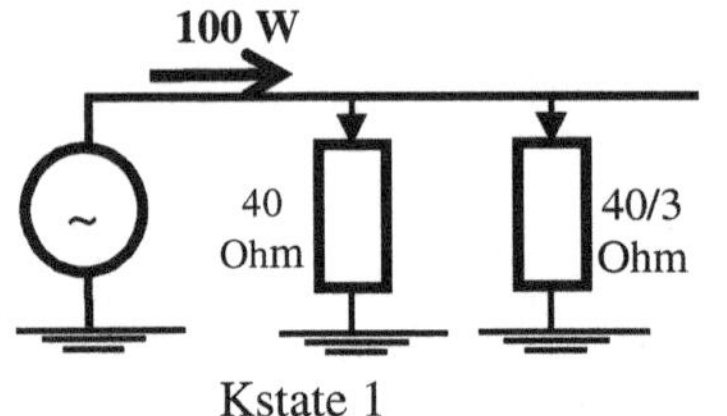

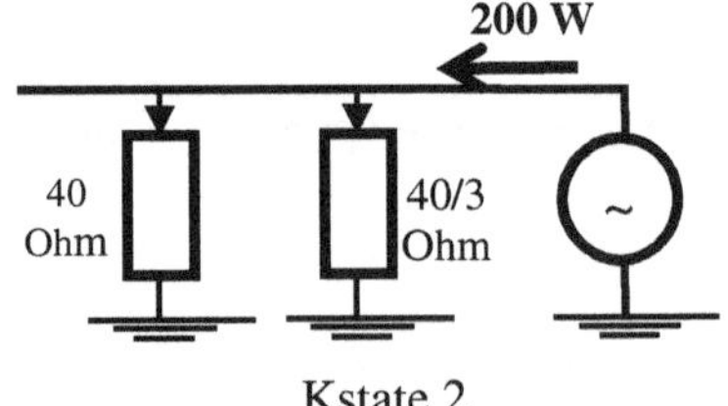

Fig. 3.2 System with Kstates

With only generator 2 (200 W):

$$
\begin{aligned}
I_{22} &= \sqrt{\frac{200}{10}} = \sqrt{20}\,\mathrm{A} \\
V_{22} &= I_{22} * R = 10\sqrt{20} \\
v_{40,2} &= v_{(40/3),2} = V_{22} = 10\sqrt{20} \\
i_{40,2} &= y_{40} v_{e2} = \frac{1}{40} * (10\sqrt{20}) = \underline{1.118}\,\mathrm{A}; \\
p_{40,2} &= i^2_{(40),2}(40) = 50\,\mathrm{W} \\
i_{(40/3),2} &= y_{40/3} v_{(40/3),2} = \frac{3}{40} * (10\sqrt{20}) = \underline{3.3541}\,\mathrm{A}; \\
p_{(40/3),2} &= i^2_{(40/3),2}(40/3) = 150\,\mathrm{W} \\
\textit{Total current with source 2 alone} &= i_{40,2} + i_{(40/3),2} = 4.459\,\mathrm{A}
\end{aligned}
$$

The two Kstates in this case are then sets,

$$\left\{\left(10\sqrt{10}, 0.7906\right), \left(10\sqrt{10}, 2.3717\right)\right\} \text{ and } \left\{\left(10\sqrt{20}, 1.118\right), \left(10\sqrt{10}, 3.3541\right)\right\}.$$

Kpowers from the two sources in 40 Ω are found to be 25 and 50 W; and those in 40/3 Ω are 75 and 150 W. Total power in 40 Ω is obtained by summing 25 and 50 W (=75 W). The total in 40/3 Ω is 225 W. Net power consumed in loads is 300 W. The net current is obtained as follows.

Voltage across 40|| (40/3) with both sources: $V(both\,sources) = \sqrt{p_e r_e} = \sqrt{300 * 10} = 54.77\,\mathrm{V}$. With net voltage of 54.77, currents in 40 and 40/3 Ω are obtained as,

$$I_{40} = \frac{54.77}{40} = 1.3693\ \text{A}$$
$$I_{\frac{40}{3}} = \frac{54.77}{40/3} = 4.1079\ \text{A}$$

Using Ohm's law,

$$\textit{Load current with both sources} = \frac{54.77}{10} = 5.477\ \text{A}$$

We observe that, $1.3693 \neq 0.7906 + 1.118$ (as also, $4.1079 \neq 2.3717 + 3.3541$) i.e., *we cannot sum Kcurrents from the two power sources to obtain total current in an element!* Superposition of Kcurrents by direct summation appears to be invalid and poses a challenge!

3.4 The Superposition Challenge

Kpowers in each element are scalars and can be summed directly. Total power in 40 Ω = 25 + 50 = 75 W; and total power in 40/3 Ω = 75 + 150 = 225 W. Superposition of total of the two element powers gives 300 W which indeed is the total power supplied as also the total consumption in the elements. It is interesting to note that the principle of superposition by summation does not apply to Kcurrents! Relation between *Kcurrents* and *net current* is observed to be,

$$1.3693^2 = 0.7906^2 + 1.118^2 \tag{3.7}$$

$$4.1079^2 = 2.3717^2 + 3.3541^2! \tag{3.8}$$

This appears to be the new principle of superposition for Kcurrents pertaining to power sources. Similar observation holds for Kvoltages. *Students are often told that superposition applies to currents in linear circuits. Relation* (3.7) *and* (3.8) *therefore unfolds interesting behaviour of voltage/currents in power-sourced circuits.* We shall analyse this in detail in Chap. 10.

3.5 Quantum Perspective

Physically, one can view each Kstate as one of the *many parallel worlds* reminding us of quantum theory. Particle-wave duals appear as Kpowers; they can resonate or annihilate. For example, Kpowers may assume opposite signs when they come into the circuit element from opposite sides, e.g., 1 Ω element in Fig. 3.3. The Kpower $p_{1\,\Omega,1}$ is positive (node 1 to node 2) when power of generator 1 flows node 1 to 2

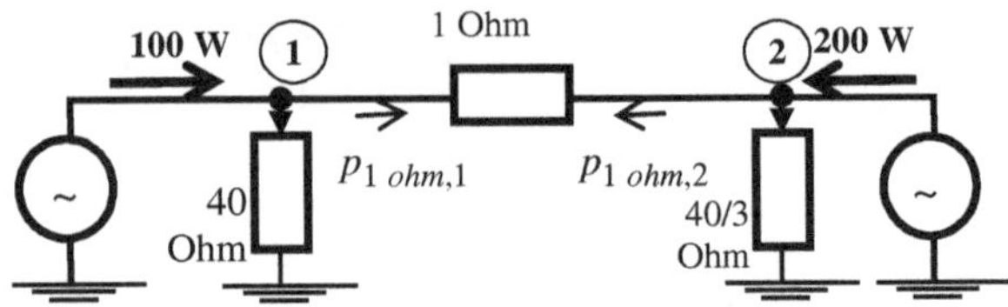

Fig. 3.3 Annihilation

and that of generator 2 negative when it flows from 2 to 1. Value of superposed power in this case is equal to, $p_{1\ \Omega,1} \sim p_{1\ \Omega,2}$ (annihilation).

Many transmission lines carry power from generators to loads in same direction (resonate) but in a few, contributions from different generators may oppose (annihilate). Such resonance or annihilation affects the analysis but little in power networks which have transmission line impedances much smaller than the load impedances.

Reflections

The idea in this chapter is an outcome of our rethink on principle of superposition. We connect one generator to the network at a time with the intention of superposing their results. Solving the network for one generator yields an electrical condition which is not physically observable since in practice all generators act simultaneously. Superposing power values in elements was no problem. Doing so with voltages/currents was. These could not be summed directly! Their superposition is observed to be an interesting revelation and is posed as a challenge in this chapter and solved in the book later.

Reference

1. H. Narayanan, Theory of Matroids, PhD Thesis, Elec. Dept, Indian Institute of Technology Bombay, 1974 (www.ee.iitb.ac.in, available on Author's Home Page)

Chapter 4
Modular Load Flow

> *.. the lofty goal is the ultimate missing component of power flow study puzzle, namely, the closed form solution.*
>
> Prof. G.T. Heydt [1]

In conventional load flow formulation, static load flow equations are set up in terms of nodal voltages. They are then solved using iterative techniques. Modular Load Flow (MLF) is formulated in terms of element voltages and makes a departure from the conventional load flow at the outset. Electrical variables associated with elements in *Kstates* (Kvoltages and Kcurrents) are the fundamental building blocks of our *modules*. We reiterate that only one generator is connected to the network in a Ksystem. When two or more generators are connected, Kstates for all Ksystems are determined separately and the net power in element is obtained by direct summation of Kpowers. The net currents are obtained by a different principle of superposition. Element voltages and currents can alternatively be calculated from net powers.

4.1 Concept of Modularity

Let's consider the network shown in Fig. 4.1. Let $P_1 = 100\,\text{W}$; $P_2 = 200\,\text{W}$; $R_1 = 1$, $R_2 = 2$, $R_3 = 99\,\Omega$. Using procedure explained in Sect. 3.3.1, the Kstate KS_1 i.e., the set of element voltages and element currents, with only generator 1, is obtained as,

$$\begin{aligned}\{v_{e1}, i_{e1}\} &= \{(v_{11}, i_{11}), (v_{21}, i_{21}), (v_{31}, i_{31})\} \\ &= \{(1,1), (0,0), (99,1)\}\end{aligned} \tag{4.1}$$

Powers in elements are,

$$\{p_{11}, p_{21}, p_{31}\} = \{1, 0, 99\} \tag{4.2}$$

M.V. Hariharan et al., *Modular Load Flow for Restructured Power Systems*, Lecture Notes in Electrical Engineering 374, DOI 10.1007/978-981-10-0497-1_4

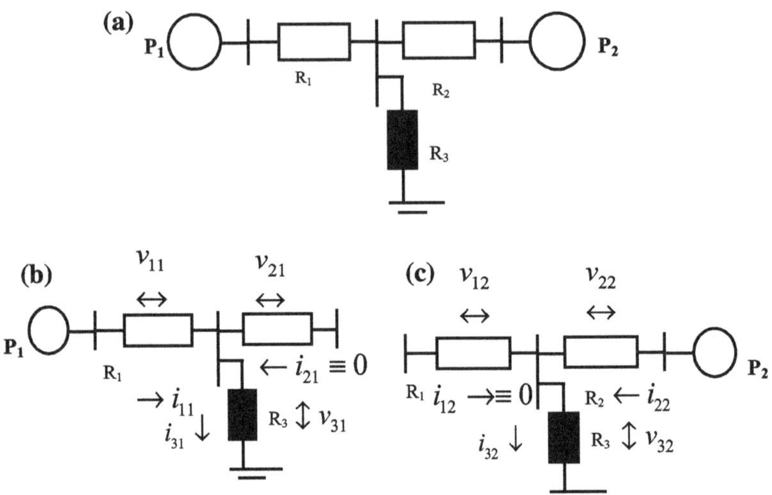

Fig. 4.1 A simple system. **a** Full system. **b** Kirchoff State: KS_1. **c** Kirchoff State 2: KS_2

Similarly, Kstate KS_2 for generator 2 is,

$$\begin{aligned}\{v_{e2}, i_{e2}\} &= \{(v_{12}, i_{12}), (v_{22}, i_{22}), (v_{32}, i_{32})\} \\ &= \{(0, 0), (2.8144, 1.4072), (139.3123, 1.4072)\}\end{aligned} \tag{4.3}$$

Element powers are,

$$\{p_{12}, p_{22}, p_{32}\} = \{0, 3.96, 196.04\} \tag{4.4}$$

A matrix equation is now formed for element powers with each row having fractions of the two generator powers in that element as shown in (4.5).

$$\begin{bmatrix} p_{R1} \\ p_{R2} \\ p_{R3} \end{bmatrix} = \left[\begin{array}{c|c} \left(\frac{1}{100}\right) & 0 \\ \hline 0 & \left(\frac{2}{101}\right) \\ \hline \left(\frac{99}{100}\right) & \left(\frac{99}{101}\right) \end{array}\right] \left[\begin{array}{c} P_{g1} = 100 \\ \hline P_{g2} = 200 \end{array}\right] \tag{4.5}$$

Equation (4.5) shows that a decoupling exists among generator powers that get distributed in circuit elements in specific proportions. For example, 100 W gets distributed in R1 and R3 in proportion 1/100:99/100, and 200 W gets distributed in elements R2 and R3 in proportion 2/101:99/101 These fractions are unaltered even if a third (or further) power source were to be connected. The premise for Modular Load Flow is thus discovered and can be stated as, *the quantum of power source gets fractioned into network elements in definite proportions, and that these power fractions can be determined from the parameters and structure of the network alone.* These fractions will be called *power fractions*. Power fractions are a *circuit property* and not functions of voltages and currents. It is good to remind ourselves that we are talking of power consumptions and not power flows. We will associate a *module* consisting of its voltage, current and power with every element. The *module* for *e*th element with respect to generator *i* is denoted by,

$$\Lambda_{ei} \triangleq \{p_{ei}, y_e, A\} \tag{4.6}$$

p_{ei} is element power; y_e is admittance of element, and A is node-element incidence matrix. Electrical description of Λ_{ei} is,

$$E_{ei} = \{v_{ei}, i_{ei}, p_{ei}\} \tag{4.7}$$

Both expressions (4.6) and (4.7) will be referred to as modules, (4.6) being understood as *symbol* and (4.7) as its *electrical description*. The *special* module for generator port is given by,

$$E_i \triangleq \{V_{ii}, I_{ii}, P_i\} \tag{4.8}$$

$$I_{ii} = \sqrt{\frac{P_i}{R_{ii}}};\ V_{ii} = Z_{ii}I_{ii} \tag{4.9}$$

There are 3 elements and 2 generators in the above example. Their modules with only source 1 are,

$$\begin{aligned} E_{11} &= \{v_{11}, i_{11}, p_{11}\} = \{1, 1, 1\} \\ E_{21} &= \{v_{21}, i_{21}, p_{21}\} = \{0, 0, 0\} \\ E_{31} &= \{(v_{31}, i_{31}, p_{31})\} = \{99, 1, 99\} \end{aligned} \tag{4.10}$$

as the element powers with only source 1 are 1, 0 and 99 in R1, R2 and R3 respectively. The Kstate for source 1 is,

$$\begin{aligned} KS_1 &= \{v_{e1}, i_{e1}\}, e = 1, 2, 3 \Rightarrow \{(v_{11}, i_{11}), (v_{21}, i_{21}), (v_{31}, i_{31})\} \\ &= \{(1, 1), (0, 0), (99, 1)\} \end{aligned} \tag{4.11}$$

Similarly, for source 2 the modules are,

$$\begin{aligned} E_{12} &= \{v_{12}, i_{12}, p_{12}\} = \{0, 0, 0\} \\ E_{22} &= \{(v_{22}, i_{22}, p_{22}\} = \{2.8144, 1.4072, 3.9604\} \\ E_{32} &= \{v_{32}, i_{32}, p_{32}\} = \{139.3123, 1.4072, 196.04\} \end{aligned} \tag{4.12}$$

And the Kstate for source 2 is,

$$\begin{aligned} KS_2 &= \{v_{e2}, i_{e2}\}, e = 1,2,3 \Rightarrow \{(v_{12}, i_{12}), (v_{22}, i_{22}), (v_{32}, i_{32})\} \\ &= \{(0,0), (2.8144, 1.4072), (139.3123, 1.4072)\} \end{aligned} \tag{4.13}$$

4.2 Power Distribution in Kstates

Using power fractions given by elements of the matrix (4.5) and the modules described in Sect. 4.1 we can visualize distribution of the generator power in a network of three elements as illustrated in Fig. 4.2. Powers p_{ei} are consumptions in the elements. In load elements it constitutes the load consumption and in line elements, it constitutes loss. With more than one source, the total loss in element e is given by Eq. (4.14)

$$p_e = sup \sum_i \varepsilon_{ei} P_i \tag{4.14}$$

Operator $sup \sum$ implies superposition; it means that loss in an element can be less than the absolute sum of individual generator powers when some generator powers flow in opposite directions in an element.

4.3 Line Flows

The node-element incidence matrix has only '+1' and '−1' elements in each column and can be written as,

$$A = A_1 - A_2 \tag{4.15}$$

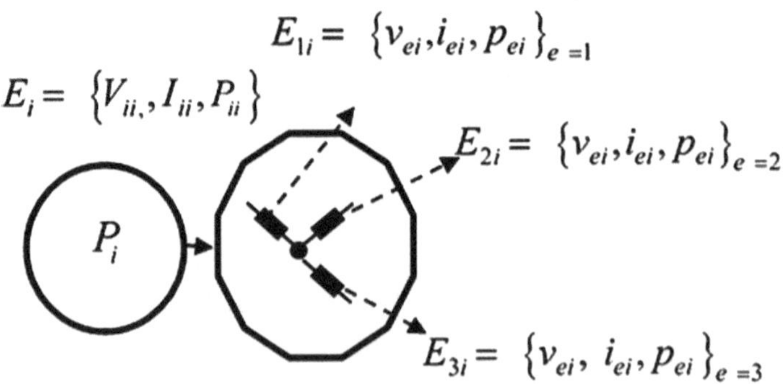

Fig. 4.2 Generator and element modules

Matrices A_1 and A_2 contain '1' in locations (*m, e*) and (*n, e*), for column corresponding to element *e* with 'from' and 'to' nodes designated by, *m* and *n,* respectively. All other elements of the two matrices are zero. We can write the ξ matrix as,

$$\begin{aligned} \xi &= \xi(1) - \xi(2) \\ \xi(1) &= \left[A_1^T Z\right]; \xi(2) = \left[A_2^T Z\right] \end{aligned} \tag{4.16}$$

The power in *e* is difference of the power flowing *into* and that flowing *out* of *e*.

$$p_{ei} = \frac{P_i}{R_{ii}} \mathrm{Re}\{[\xi(1) - \xi(2)]_{ei}[\xi(1) - \xi(2)]_{ie}^* y_e^*\} \tag{4.17}$$

$$p_{ei} = \mathrm{Re}\left\{\frac{P_i}{R_{ii}} \xi(1)_{ei}[\xi(1) - \xi(2)]_{ie}^* y_e^*\right\} - \mathrm{Re}\left\{\frac{P_i}{R_{ii}} \xi(2)_{ei}[\xi(1) - \xi(2)]_{ie}^* y_e^*\right\} \tag{4.18}$$

First term is power flowing *into* the element *e*, and the other that flowing *out*, i.e.,

$$\begin{aligned} P_{eif-in} &= \frac{P_i}{R_{ii}} \mathrm{Re}\{\xi(1)_{ei}[\xi(1) - \xi(2)]_{ie}^* y_e^*\} \\ P_{eif-in} &= \varepsilon_{eif} P_i \end{aligned} \tag{4.19}$$

$$\varepsilon_{eif} \triangleq \mathrm{Re}\left\{\frac{1}{R_{ii}} \left(\xi(1)_{ei} \xi_{ie}^* y_e^*\right)\right\} \tag{4.20}$$

Power flow *on* line *e* due to generator *i* is assumed to be equal to that flowing *into* it,

$$P_{eif} = \varepsilon_{eif} P_i \tag{4.21}$$

Total flow on line *e* is then,

$$P_{ef} = sup \sum_i \varepsilon_{eif} P_i \tag{4.22}$$

It may be recalled that power loss in line *e* due to generator *i* is given by (3.3), reproduced here for ready reference,

$$\begin{aligned} p_{ei} &= \frac{P_i}{R_{ii}} \mathrm{Re}\{|\xi_{ei}|\xi_{ie}^* y_e^*\} \\ &= \varepsilon_{ei} P_i \end{aligned} \tag{4.23}$$

And the total loss by, Eq. (4.14) reproduced

$$p_e = sup \sum_i \varepsilon_{ei} P_i \tag{4.14}$$

4.4 Superposition of Line Flows

It is to be noted that we obtained powers flowing through elements without explicitly calculating voltages. We also did not use *sum of the currents-squared-resistance* formula to find power-loss in the element. We obtained them from fractional powers of individual generators in elements. Superposition of *powers* has been an anathema to power community. Here we wish to point out that a fundamental property of the superposition principle is usually not recognized—that it applies to the variable of forcing function; it applies to current variables in circuits with voltage/current sources, and to powers in power-sourced circuits. Currents cannot be superposed in power-sourced circuits just as powers cannot be superposed in voltage sourced circuits. Currents become nonlinear functions (square roots) of powers, just as powers are nonlinear functions of voltage/currents! We will discuss more of this in Chap. 10.

4.5 Voltages

Element voltage for a resistive circuit is also obtained in terms of power as,

$$v_e = \sqrt{p_e r_e} \tag{4.24}$$

Or, from the element-voltage triangle in an ac circuit (Fig. 4.3),

$$v_e = \sqrt{p_e r_e + q_e r_e} = \sqrt{p_e r_e\left(1 + \frac{x_e^2}{r_e^2}\right)} \tag{4.25}$$

4.6 Phase Angles

In order to calculate phase angle difference δ_e between two adjacent buses, we make use of the voltage triangle formed by magnitudes of the two adjacent bus voltages V_1 *and* V_2 (voltages across the load impedances connected to these buses)

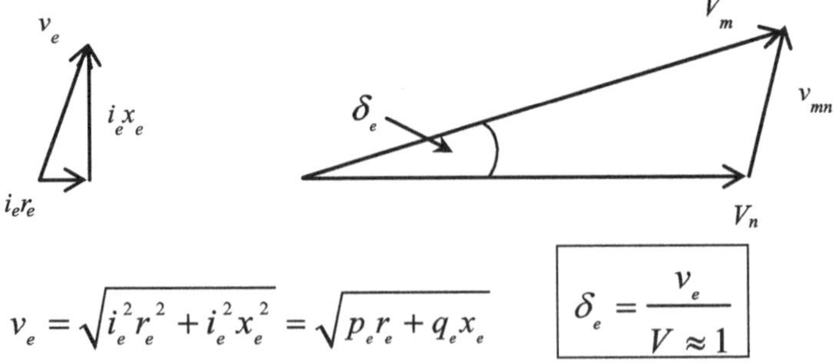

Fig. 4.3 Element-voltage and node-voltage triangles

and, the voltage of the element between the two buses, by the cosine formula (see Fig. 4.3),

$$\cos \delta_e = \frac{V_1^2 + V_2^2 - v_e^2}{2V_1 V_2} \tag{4.26}$$

With assumptions, $V_1 \approx V_2 \approx V \approx 1$,

$$\cos \delta_e = 1 - \frac{v_e^2}{2} \tag{4.27}$$

Representing LHS by approximated series expansion of cosine function and comparing the two sides,

$$\delta_e = v_e \tag{4.28}$$

The value of element voltage thus is also equal to the phase angle difference, in radians, between two of its end buses!

4.7 Modular Load Flow Algorithm

Based on the analysis given above we can now state an algorithm to find line-flows, losses, voltages and phase-angle differences across elements in a power network.

Algorithm

1. Convert load P and Q into equivalent impedances (at nominal voltages).
2. Convert all Q injections (shunt reactors, synchronous condensers and synchronous generators) into shunt reactances (admittances). Use nominal 1 pu voltage if actual voltages are not available.
3. Include impedances obtained in steps 2 and 3 in the network as elements, and construct node-element incidence matrix A, primitive admittance matrix [y], Y-bus, bus impedance matrix Z-bus and the ξ matrix.
4. Obtain power flows in elements using (4.19)–(4.22).
5. Obtain power loss in elements using (4.23) and (4.14) reproduced after (4.23).
6. Obtain voltages and phase angle difference across each using (4.25) and (4.28).

In Chap. 5 we will solve 3-bus, IEEE 30-bus and an ill-conditioned 43-bus examples using above algorithm. Below we compare computational sequence in MLF with that in the ILF. ILF procedure first reported in fifties has limitations especially in solving the ill-conditioned systems.

4.8 Comparison with Conventional Load Flow

Computational sequence in ILF and MLF are shown in Fig. 4.4.

In ILF two variables must be specified apriori at *every* bus before proceeding with the Gaus-Siedel or Newton-Rapson iterative procedure. In MLF we need only *ng* real powers and *ng* reactive powers injected by the generators into the network, and their voltages. Compare this against apriori specifications equal to twice the number of *buses* in an ILF. When real-time injections are used for calculations, more reliable values of losses and flows are obtained by MLF. Recall that phase angles are a prerequisite for calculation of power losses/flows in ILF. Absence of phase angles in MLF may appear to be strange but power flows can indeed be obtained without phase angle information using (4.19)–(4.22). Phase angle differences across elements, if at all required, are computed from (4.27/4.28). Procedural accuracy is guaranteed because of the closed-form algebraic expressions. Convergence in ILF depends on the initial guess but is no longer a problem in MLF. Solution is assured. MLF has been found to be 6–10 times faster than the ILF. Assigning all losses to the slack generator in ILF is a mathematical necessity, but is a defective practice. It has no physical relevance. Concept of slack generator is not employed at all in MLF. Power balance is the foundation of modular approach and is maintained at every stage. Remarkable feature of MLF is its ability to determine individual generator contributions which have commercial implications. ILF does not provide this information. With ILF, distribution factors and tracing are required as *add-ons* for this purpose [2, 3]. Both procedures have limitations. MLF can be used when quick and accurate computation of line flows, losses and *magnitudes* of voltages is required. That the phase angle is not required for power computation is a novel and interesting feature of MLF.

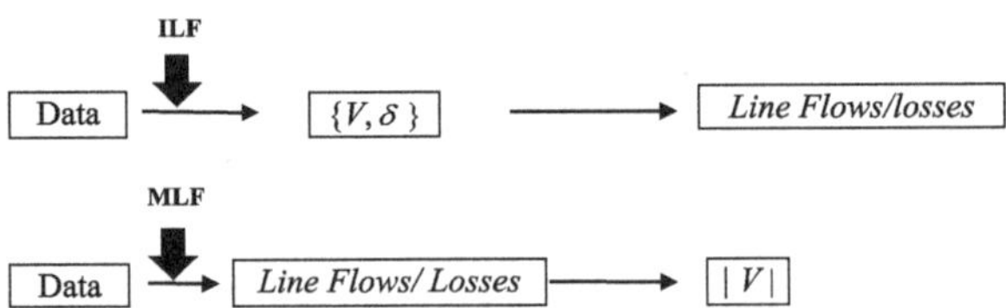

Fig. 4.4 Sequence of computations in ILF and MLF

4.9 HVDC, FACTS and DSG

New technologies have always created need for extending application of existing analytical methods. More and more insight is gained when applications mature. Advancements in high voltage DC transmission (HVDC), flexible AC transmission (FACTS) and dispersed system generation (DSG) naturally led to research in iterative load flow methodologies. At this point of time MLF is in its infancy and authors are not aware of any publications, let alone practical applications of the concept. Once it is accepted by professionals, these areas could be explored further. Prima-facie, we do not see any reason why such extensions should not be possible. For example, an HVDC line can be replaced by an appropriate load at one end, and, injection at the other. All other analytical steps would remain same. FACTS can be simulated by appropriate tuning of line reactances. It appears a little daunting to be able to determine sensitivity of the ξ matrix with respect to line parameters, but Z-bus building algorithm can resolve this issue to some extent. Distributed generations are injections of relatively smaller amounts of powers in a huge system. Intermittency of DSGs would cause ripples in electrical variables rather than actuate load-generation balancing mechanism by governors. We believe this to be an appropriate area for application of MLF. One such application is reported in [4].

Reflections

While working with simple examples and trying to look for relation between element powers and the injected powers we noticed that, in singly excited network there was always a constant proportion between the two powers! How could this ratio be a constant? With different injections, the currents and voltages in elements would change in a manner not fully known! If the observations were true we must have an analytical proof! This chapter discovers it in the formulation. Neat analytical expressions are obtained for losses and flows! What a surprise, when we look back as to how we ourselves and many researchers in the past had struggled to separate the loss/flow contributions of generators for allocation purposes by the Independent System Operators (ISO) in deregulated systems! The thought that these can be accurately and explicitly determined is a great relief! Efforts to understand the mechanism of superposition of currents and voltages led to a picture of orthogonal spaces which we found very exciting. This chapter summarizes initial formulation for this purpose. The picture part will be reported in Chap. 10. *In summary, superposition of power in power-sourced circuits works brilliantly; currents and voltages follow a new mechanism of superposition.*

References

1. G.T. Heydt, Disc. on, On formulation of power distribution factors for linear load flow problem. IEEE Trans. Power Apparatus and Syst. (P. W. Sauer) **PAS-100**(3), 558 (1981)
2. A.J. Wood, B.F. Wollenberg, *Power Generation, Operation and Control*, 2nd edn. (John Wiley and Sons, New York, 1996)
3. J. Bialek, Tracing the flow of electricity. IEE Proc.-Gen Trans. Distrib. **143**, 313–320 (1996)
4. S.D. Varwandkar, M.V. Hariharan, Plugin Power Flow, in *6th IEEE Power India International Conference* (PIICON-2014), New Delhi, India, 5–7 Dec 2014

Chapter 5
Load Flow Examples

The proof of the pudding is in the eating ...

In this chapter we present results on three typical power systems. First is a basic 3-bus system. Reader should be able to check our results easily. The other two are the IEEE 30-bus system [1], and a 43-bus system [2, 3]. While the IEEE 30-bus system is a well-conditioned system and does not pose convergence issues with conventional Newton–Raphson load flow, the 43-bus system is an ill-conditioned one with poor condition number. It incorporates features of radial distribution systems which call for specially devised versions of Newton–Raphson method. The system has engaged many researchers for its adamant non-convergence [2–4]. MLF is a one-shot procedure and with its attendant advantage is equally effective for both types of systems. Line parameters for 43-bus system were extracted from the off-diagonal elements of the given Y-bus matrix [3] (Y-bus is the only form in which line data is available). Extracting line charging and transformer tap settings was not possible. Half-line charging admittance of 0.0001 pu was added to each line. Y-bus with this line charging is found to be close to the given Y-bus. The modifications are therefore considered marginal and it is possible to draw general conclusions from results.

5.1 The 3-Bus System

Figure 5.1 shows the system. Data in pu are given on the figure.

Half line charging for lines: Line 1–3 = 0.001; Line 2–3 = 0.0001; Line 1–2 = 0.0001.

5.1.1 Results for 3-Bus System

Primitive admittance matrix for system shown in Fig. 5.1 is a diagonal matrix with elements; three for lines, three for total charging admittances at the three buses, the

M.V. Hariharan et al., *Modular Load Flow for Restructured Power Systems*,
Lecture Notes in Electrical Engineering 374, DOI 10.1007/978-981-10-0497-1_5

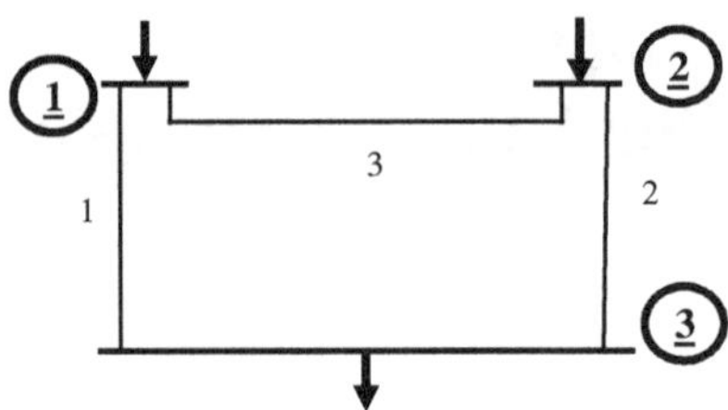

Fig. 5.1 3 Bus system

load admittance, and two admittances representing generator reactive power injections. These diagonal elements are,

$0.3846 - j1.9231$ (line 1–3); $0.1923 - j0.9615$ (line 2–3); $0.5405 - j3.2432$ (line 1–2); $j0.0011$ (net charging admittance @ 1); $j0.0002$ (net charging admittance @ 2); $j0.0011$ (net charging admittance @ 3); $0.5 - j0.2$ (load admittance); $j0.2161$; $j0.103$ (generator Qs are converted to shunt admittances)—total nine values.

The bus incidence matrix is,

$$A = \begin{bmatrix} 1 & 0 & 1 & 1 & 0 & 0 & 0 & 1 & 0 \\ 0 & 1 & -1 & 0 & 1 & 0 & 0 & 0 & 1 \\ -1 & -1 & 0 & 0 & 0 & 1 & 1 & 0 & 0 \end{bmatrix}$$

Bus admittance matrix is,,

$$Y = \begin{bmatrix} 0.9254 - \mathrm{j}4.9491 & -0.5405 + \mathrm{j}3.2432 & -0.3846 + \mathrm{j}1.9231 \\ -0.5405 + \mathrm{j}3.2432 & 0.7329 - \mathrm{j}4.1016 & -0.1923 + \mathrm{j}0.9615 \\ -0.3846 + \mathrm{j}1.9231 & -0.1923 + \mathrm{j}0.9615 & 1.0769 - \mathrm{j}3.0835 \end{bmatrix}$$

$$Z = \begin{bmatrix} 2.2943 - j0.4105 & 2.2775 - j0.4945 & 1.987 - j0.6760 \\ 2.2775 - j0.4945 & 2.3052 - j0.3242 & 1.9853 - j0.6750 \\ 1.987 - j0.6760 & 1.9853 - j0.6750 & 1.7889 - j0.5609 \end{bmatrix}$$

The ξ matrix is,

$$\xi = \left[\begin{array}{c|c} 0.3073 + j0.2654 & 0.29213 + j0.18052 \\ 0.2904 + j0.1814 & 0.31984 + j0.35086 \\ 0.01689 + j0.0840 & -0.02771 - j0.17034 \\ 2.2943 - j0.41053 & 2.2775 - j0.49454 \\ 2.2775 - j0.49454 & 2.3052 - j0.3242 \\ 1.987 - j0.67601 & 1.9853 - j0.67507 \\ 1.987 - j0.67601 & 1.9853 - j0.67507 \\ 2.2943 - j0.41053 & 2.2775 - j0.49454 \\ 2.2775 - j0.49454 & 2.3052 - j0.3242 \end{array}\right]$$

Line flows (Table 5.1) are obtained from (4.19) and (4.22).

Bus voltages (voltage across impedances connected to these buses) are obtained from (4.25) (Table 5.2).

Table 5.1 Line Flows in elements (pu)

From	To	Flow contributions (pu)		Total flow
		G1	G2	
1	3	0.2322	0.1119	0.3441
2	3	0.0912	0.0865	0.1777
1	2	0.0918	−0.1120	−0.0202

Table 5.2 Bus voltages (pu)

Bus #	1	2	3
Voltage	1.1131	1.1125	1.002

5.2 The IEEE 30-Bus System [1]

See Fig. 5.2 and Tables 5.3 and 5.4

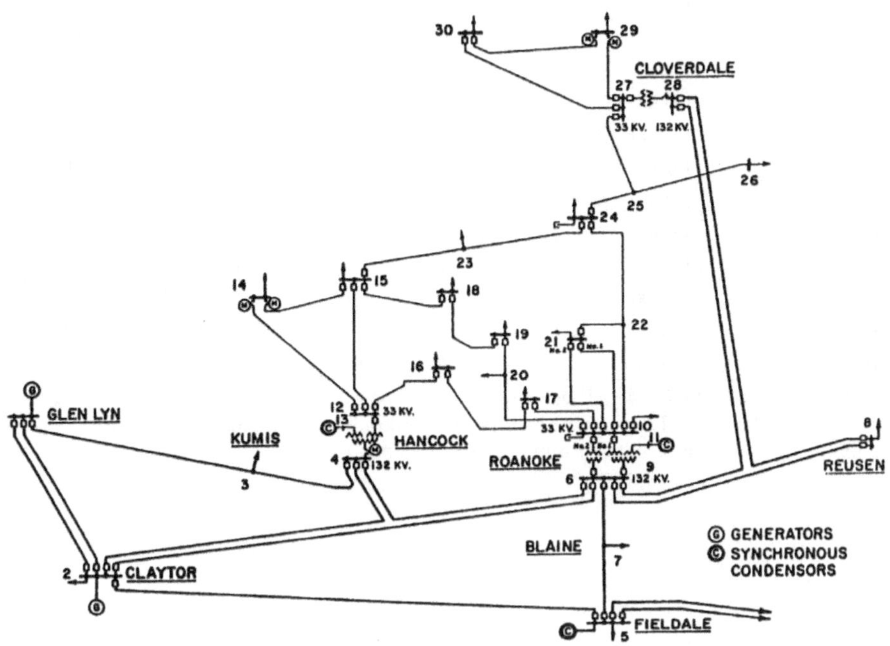

Fig. 5.2 IEEE 30 bus system

Table 5.3 Bus data (voltage (pu); powers (MW))

Bus	Type	Voltage	Angle	Pg	Qg	PL	QL
1	1	1.0600	0	0[a]	0[a]	0	0
2	2	1.0430	0	40	50.0000	21.7000	12.7000
3	3	1.0000	0	0	0	2.4000	1.2000
4	3	1.0600	0	0	0	7.6000	1.6000
5	2	1.0100	0	0	37.0000	94.2000	19.0000
6	3	1.0000	0	0	0	0	0
7	3	1.0000	0	0	0	22.8000	10.9000
8	2	1.0100	0	0	37.3000	30.0000	30.0000
9	3	1.0000	0	0	0	0	0
10	3	1.0000	0	0	19.0000	5.8000	2.0000
11	2	1.0820	0	0	16.2000	0	0
12	3	1.0000	0	0	0	11.2000	7.5000
13	2	1.0710	0	0	10.6000	0	0
14	3	1.0000	0	0	0	6.2000	1.6000
15	3	1.0000	0	0	0	8.2000	2.5000
16	3	1.0000	0	0	0	3.5000	1.8000
17	3	1.0000	0	0	0	9.0000	5.8000
18	3	1.0000	0	0	0	3.2000	0.9000
19	3	1.0000	0	0	0	9.5000	3.4000
20	3	1.0000	0	0	0	2.2000	0.7000
21	3	1.0000	0	0	0	17.5000	11.2000
22	3	1.0000	0	0	0	0	0
23	3	1.0000	0	0	0	3.2000	1.6000
24	3	1.0000	0	0	4.3000	8.7000	6.7000
25	3	1.0000	0	0	0	0	0
26	3	1.0000	0	0	0	3.5000	2.3000
27	3	1.0000	0	0	0	0	0
28	3	1.0000	0	0	0	0	0
29	3	1.0000	0	0	0	2.4000	0.9000
30	3	1.0000	0	0	0	10.6000	1.9000

[a]261.051 MW and −14.70 MVAR are used at bus 1, and 40 MW and 54.536 MVAR at bus 2 to perform Modular Load Flow. These values are purposely chosen from the output of an NR loadflow to get an idea of difference in results by the two methods

Table 5.4 Line data (pu)

From Bus	To Bus	R	X	B/2	Tap
1	2	0.0192	0.0575	0.0264	1.0000
1	3	0.0452	0.1652	0.0204	1.0000
2	4	0.0570	0.1737	0.0184	1.0000
3	4	0.0132	0.0379	0.0042	1.0000
2	5	0.0472	0.1983	0.0209	1.0000
2	6	0.0581	0.1763	0.0187	1.0000
4	6	0.0119	0.0414	0.0045	1.0000
5	7	0.0460	0.1160	0.0102	1.0000
6	7	0.0267	0.0820	0.0085	1.0000
6	8	0.0120	0.0420	0.0045	1.0000
6	9	0	0.2080	0.0001	0.9780
6	10	0	0.5560	0.0001	0.9690
9	11	0	0.2080	0.0001	1.0000
9	10	0	0.1100	0.0000	1.0000
4	12	0	0.2560	0.0000	0.9320
12	13	0	0.1400	0.0000	1.0000
12	14	0.1231	0.2559	0.0000	1.0000
12	15	0.0662	0.1304	0.0000	1.0000
12	16	0.0945	0.1987	0.0000	1.0000
14	15	0.2210	0.1997	0.0000	1.0000
16	17	0.0824	0.1923	0.0000	1.0000
15	18	0.1073	0.2185	0.0000	1.0000
18	19	0.0639	0.1292	0.0000	1.0000
19	20	0.0340	0.0680	0.0000	1.0000
10	20	0.0936	0.2090	0.0000	1.0000
10	17	0.0324	0.0845	0.0000	1.0000
10	21	0.0348	0.0749	0.0000	1.0000
10	22	0.0727	0.1499	0.0000	1.0000
21	23	0.0116	0.0236	0.0000	1.0000
15	23	0.1000	0.2020	0.0000	1.0000
22	24	0.1150	0.1790	0.0000	1.0000
23	24	0.1320	0.2700	0.0000	1.0000
24	25	0.1885	0.3292	0.0000	1.0000
25	26	0.2544	0.3800	0.0000	1.0000
25	27	0.1093	0.2087	0.0000	1.0000
28	27	0	0.3960	0.0000	0.9680
27	29	0.2198	0.4153	0.0000	1.0000
27	30	0.3202	0.6027	0.0000	1.0000
29	30	0.2399	0.4533	0.0000	1.0000
8	28	0.0636	0.2000	0.0214	1.0000
6	28	0.0169	0.0599	0.0065	1.0000

5.2.1 Results for 30-Bus System

See Tables 5.5 and 5.6.

Table 5.5 Line flows with individual generator contributions

From	To	Flow contributions (MW)		Total Flow
		G1	G2	
1	2	180.2153	−5.7361	174.4792
1	3	80.8183	5.7332	86.5515
2	4	34.3323	8.6053	42.9376
3	4	75.6657	5.3038	80.9695
2	5	70.9249	12.3128	83.2377
2	6	49.7050	10.2373	59.9423
4	6	65.4628	7.1249	72.5877
5	7	−13.6714	−1.1504	−14.8219
6	7	34.0110	4.3177	38.3287
6	8	25.4208	3.9852	29.4060
6	9	23.8075	3.8450	27.6525
6	10	13.5998	2.1964	15.7962
9	11	0.0196	0.0031	0.0227
9	10	23.7786	3.8405	27.6191
4	12	36.3032	5.4346	41.7378
12	13	0.0179	0.0028	0.0207
12	14	6.3346	0.9596	7.2942
12	15	14.4972	2.1504	16.6476
12	16	6.1445	0.8705	7.0150
14	15	1.1035	0.1443	1.2478
16	17	3.1658	0.4066	3.5724
15	18	5.1627	0.7641	5.9269
18	19	2.4340	0.3385	2.7724
19	20	−5.5781	−0.9133	−6.4914
10	20	7.5313	1.2194	8.7507
10	17	4.4654	0.7860	5.2514
10	21	15.6397	2.5125	18.1522
10	22	4.8121	0.7482	5.5603
21	23	0.7066	0.1773	0.8839
15	23	3.4111	0.4362	3.8474

(continued)

Table 5.5 (continued)

From	To	Flow contributions (MW)		Total Flow
		G1	G2	
22	24	4.7791	0.7430	5.5221
23	24	1.3945	0.1886	1.5831
24	25	−1.2544	−0.2297	−1.4841
25	26	3.0261	0.4735	3.4996
25	27	−4.2945	−0.7056	−5.0001
28	27	15.6011	2.4757	18.0768
27	29	5.2605	0.8234	6.0839
27	30	6.0126	0.9411	6.9537
29	30	3.1284	0.4897	3.6181
[a]8	28	−0.5006	−0.0716	−0.5722
6	28	16.1631	2.5571	18.7202
2	31	19.1147	3.0545	22.1691
3	31	2.0982	0.3226	2.4208
4	31	6.5834	1.0250	7.6085
5	31	82.0821	12.9882	95.0703
7	31	19.7302	3.1017	22.8319
8	31	25.8111	4.0395	29.8507
10	31	4.9115	0.7680	5.6796
12	31	9.2998	1.4498	10.7496
14	31	5.1654	0.8056	5.9710
15	31	6.8551	1.0699	7.9251
16	31	2.9297	0.4574	3.3870
17	31	7.5945	1.1870	8.7816
18	31	2.6873	0.4197	3.1070
19	31	7.9990	1.2498	9.2489
20	31	1.8551	0.2899	2.1450
21	31	14.7804	2.3104	17.0909
23	31	2.7008	0.4221	3.1230
24	31	7.3693	1.1523	8.5216
26	31	2.9791	0.4661	3.4452
29	31	2.0499	0.3209	2.3708
30	31	8.9666	1.4035	10.3700

[a]This point onwards the elements are loads

Table 5.6 Bus voltages (pu)

Load bus	Voltage (pu)
2	1.0542
3	1.0221
4	1.0086
5	1.0147
7	1.0003
8	0.9975
10	1.0248
12	1.0279
14	1.0152
15	1.0126
16	1.0193
17	1.0178
18	1.0048
19	1.0033
20	1.0079
21	1.0090
23	1.0084
24	1.0049
26	0.9865
29	0.9925
30	0.9812

5.3 The Ill-Conditioned 43-Bus System

See Fig. 5.3 and Tables 5.7, 5.8, 5.9 and 5.10.

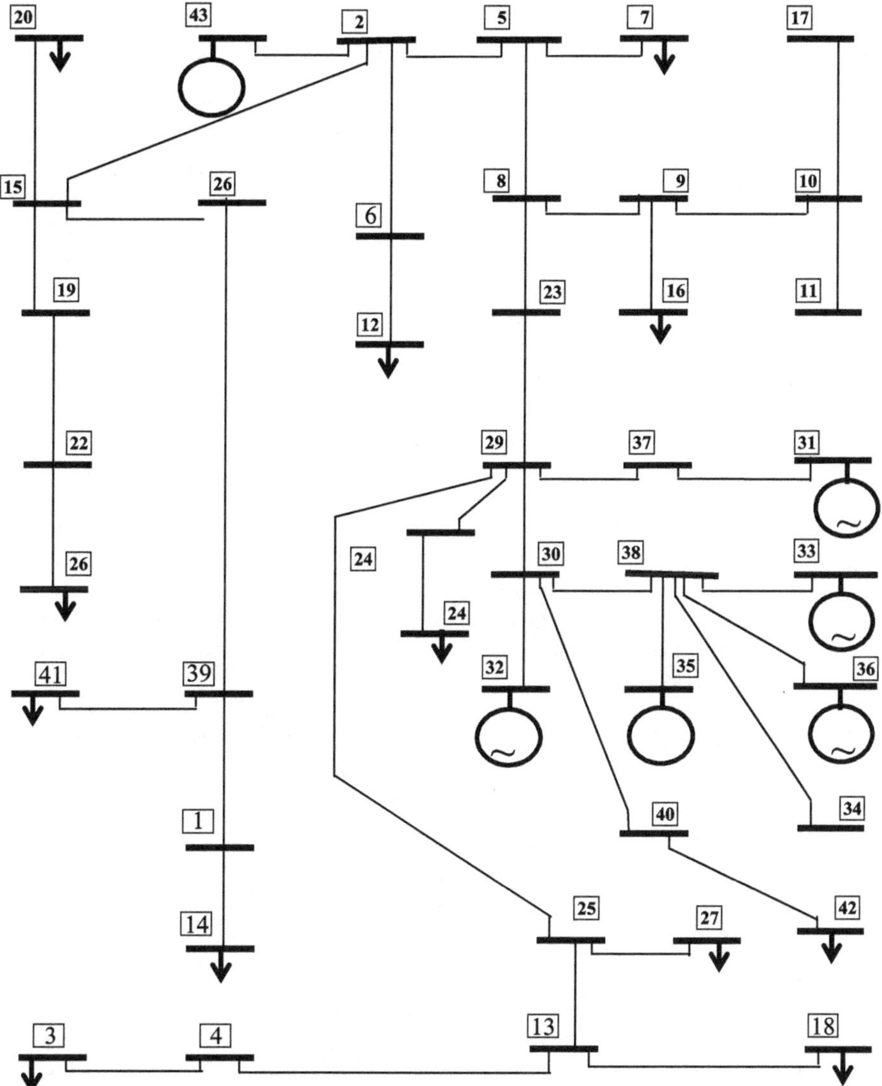

Fig. 5.3 The Ill-Conditioned 43 bus system

Table 5.7 Y-bus elements (pu)

Element subscripts		G	B	Element subscripts		G	B
1	1	309.8060	−408.0290	21	21	104.3120	−143.6090
1	14	0	15.4000	21	24	0	9.2670
1	39	−309.8060	392.2550	21	29	−104.3120	133.6230
2	2	481.2880	−1545.194	22	22	164.2920	-282.2810
2	5	−277.1950	873.5830	22	26	0	9.0230
2	6	−34.3680	108.1240	23	23	321.5790	−328.8100
2	15	−169.7260	534.3220	23	29	−157.6770	161.7600
2	43	0	30.6090	24	24	0	−8.5720
3	3	0	−5.7140	25	25	87.1500	−106.8140
3	4	0	6.0150	25	27	0	9.0230
4	4	61.3310	−69.1600	25	29	−56.1000	65.8240
4	13	−61.3310	62.8740	26	26	0	−8.5720
5	5	277.1950	−916.8920	27	27	0	−8.5720
5	7	0	21.2770	28	28	373.4470	−612.8370
5	8	0	20.5130	28	39	−202.7750	256.1360
6	6	34.3680	−118.6990	29	29	318.0890	−372.3110
6	12	0	10.6380	29	30	0	3.7660
7	7	0	−20.0000	29	37	0	7.8950
8	8	452.8400	−482.8610	30	30	125.7890	−524.4640
8	9	−288.9380	295.7770	30	32	0	30.7690
8	23	−163.9020	167.1910	30	38	0	4.1310
9	9	300.9830	−317.0440	30	40	−125.7890	485.5470
9	10	−12.0450	12.3420	31	31	0	−13.0380
9	16	0	8.7960	31	37	0	13.0380
10	10	12.0450	−20.8550	32	32	0	−30.7690
10	11	0	2.8570	33	33	0	−3.3200
10	17	0	5.7140	33	38	0	3.3200
11	11	0	−2.8570	34	34	0	−7.3650
12	12	0	−10.0000	34	38	0	6.8520
13	13	92.3810	−100.7090	35	35	0	−6.1800
13	18	0	6.0150	35	38	0	6.1800
13	25	−31.0500	31.6400	36	36	0	−2.7030
14	14	0	−15.0150	36	38	0	2.7030
15	15	340.3980	−916.7830	37	37	0	−21.3480
15	19	0	8.6490	38	38	0	−22.3980
15	20	0	15.7910	39	39	512.5810	−663.2600
15	28	−170.6730	357.0030	39	41	0	15.0150
16	16	0	−8.5760	40	40	125.7890	−508.8370
17	17	0	−5.7140	40	42	0	21.6220
18	18	0	−5.7140	41	41	0	−15.0150
19	19	164.2920	−280.7830	42	42	0	−20.0000
19	22	−164.2920	272.8050	43	43	0	−30.6090
20	20	0	−15.0020				

Table 5.8 Line data[a]

From bus	To bus	R	X
1	14	0.0001	0.0649
1	39	0.0012	0.0016
2	5	0.0003	0.0010
2	6	0.0027	0.0084
2	15	0.0005	0.0017
2	43	0.0001	0.0327
3	4	0.0001	0.1663
4	13	0.0079	0.0081
5	7	0.0001	0.0470
5	8	0.0001	0.0487
6	12	0.0001	0.0940
8	9	0.0017	0.0017
8	23	0.0030	0.0031
9	10	0.0405	0.0415
9	16	0.0001	0.1137
10	11	0.0001	0.3500
10	17	0.0001	0.1750
13	18	0.0001	0.1663
13	25	0.0158	0.0161
15	19	0.0001	0.1156
15	20	0.0001	0.0633
15	28	0.0011	0.0023
19	22	0.0016	0.0027
21	24	0.0001	0.1079
21	29	0.0036	0.0047
22	26	0.0001	0.1108
23	29	0.0031	0.0032
25	27	0.0001	0.1108
25	29	0.0075	0.0088
28	39	0.0019	0.0024
29	30	0.0001	0.2655
29	37	0.0001	0.1267
30	32	0.0001	0.0325
30	38	0.0001	0.2421
30	40	0.0005	0.0019
31	37	0.0001	0.0767
33	38	0.0001	0.3012
34	38	0.0001	0.1459
35	38	0.0001	0.1618
36	38	0.0001	0.3700
39	41	0.0001	0.0666
40	42	0.0001	0.0462

[a]Line resistance, if zero in [3], has been substituted by 0.0001. Half line charging of 0.0001 has been added

Table 5.9 Generator data

At bus	P + jQ (MW + jMVAR)
31	116.00 + j52.00
32	290.00 + j257.0
33	28.50 + j30.00
35	58.00 + j56.00
43	800.00 + j65.00

Instead of specifying voltage, we specify power at bus 43 which was named as 'slack bus' in the conventional load flow procedure. This power value is obtained from (Total load—All generation except bus 43 + "5–15 % of the total" to account for losses)

Table 5.10 Load data

Bus	P (MW)	Q (MVAR)
3	16	12
5	53	40
7	160	120
12	80	60
14	80	60
16	64	48
18	24	18
20	88	66
24	64	48
26	80	60
27	32	24
36	5	3
38	144	102
41	80	30
42	224	168

5.3.1 Results for 43-Bus System

See Tables 5.11 and 5.12.

Table 5.11 Bus voltages

Bus	Voltage (pu)	Bus	Voltage (pu)	Bus	Voltage (pu)
1	1.0218	16	1.0224	31	1.0516
2	1.0235	17	1.0281	32	1.0975
3	1.0243	18	1.0235	33	1.1078
4	1.0263	19	1.0159	34	1.0913
5	1.0233	20	1.0187	35	1.1078
6	1.0227	21	1.0283	36	1.0925
7	1.0175	22	1.0156	37	1.0423
8	1.0282	23	1.0285	38	1.0912
9	1.028	24	1.0229	39	1.022
10	1.0281	25	1.0277	40	1.0835
11	1.0281	26	1.0088	41	1.02
12	1.017	27	1.025	42	1.0751
13	1.0266	28	1.0226	43	1.0357
14	1.0178	29	1.0287		
15	1.0229	30	1.084		

Table 5.12 Line Flows

From	To	Contributions from generators (MW)					Total flow (MW)
		G1	G2	G3	G4	G5	
1	14	7.482	17.139	1.6154	3.2875	53.358	82.882
1	13	−7.482	−17.139	−1.6154	−3.2875	−53.359	-82.882
2	39	−38.084	−87.243	−8.2226	−16.734	528.28	378.0
2	5	7.4716	17.116	1.6132	3.283	53.286	82.77
2	6	30.607	70.116	6.6084	13.449	218.28	339.06
2	15	0.0048	0.0110	0.0010	0.0021	−799.85	−799.83
3	43	−1.5805	−3.6206	−0.34124	−0.694	−10.55	−16.787
4	4	−1.5805	−3.6207	−0.34125	−0.694	−10.55	−16.787
5	7	14.97	34.293	3.2321	6.5776	106.59	165.66
5	8	−58.074	−133.04	−12.539	−25.518	385.85	156.68
6	12	7.4691	17.11	1.6127	3.2819	53.268	82.742
8	9	6.2216	14.253	1.3433	2.7337	42.365	66.917
8	23	−64.301	−147.3	−13.883	−28.254	343.46	89.723
9	10	0	0	0	0	0	0
9	16	6.2204	14.25	1.3431	2.7332	42.358	66.904
10	11	0	0	0	0	0	0
10	17	0	0	0	0	0	0
13	18	2.367	5.4225	0.51107	1.0401	15.801	25.141
13	25	−3.948	−9.044	−0.85239	−1.7347	−26.353	−41.932
15	19	7.3514	16.841	1.5872	3.2302	52.429	81.439
15	20	8.2432	18.884	1.7798	3.622	58.789	91.318

(continued)

Table 5.12 (continued)

From	To	Contributions from generators (MW)					Total flow (MW)
		G1	G2	G3	G4	G5	
15	28	15.005	34.375	3.2398	6.5933	107.01	166.23
19	22	7.3513	16.841	1.5872	3.2301	52.428	81.437
21	24	6.3052	14.444	1.3614	2.7705	42.089	66.97
21	29	−6.3052	−14.444	−1.3614	−2.7705	−42.089	−66.97
22	26	7.3498	16.837	1.5869	3.2295	52.417	81.421
23	29	−64.485	−147.72	−13.923	−28.334	342.93	88.469
25	27	3.1651	7.2507	0.68338	1.3907	21.128	33.618
25	29	−7.1169	−16.304	−1.5366	−3.1271	−47.507	−75.591
28	39	15.002	34.367	3.2391	6.5917	106.99	166.19
29	30	37.856	−178.94	−16.865	−34.321	252.7	60.434
29	37	−115.96	0.0123	0.0011	0.0023	0.0359	−115.91
30	32	0.0275	−289.88	0.0076	0.0154	0.1838	−289.64
30	38	15.203	44.578	−23.127	−47.065	101.48	91.702
30	40	22.625	66.34	6.2525	12.724	151.02	258.97
31	37	115.99	−0.0121	−0.0011	−0.0023	−0.00354	115.94
33	38	−0.0031	−0.0091	28.498	−0.0022	−0.0208	28.463
34	38	0	0	0	0	0	0
35	38	−0.0058	−0.0170	−0.0020	57.995	−0.0388	57.931
36	38	−0.51	−1.4983	−0.18039	−0.36711	−3.411	−5.9678
39	41	7.51	17.211	1.6221	3.3012	53.581	83.228
40	42	22.621	66.328	6.2514	12.722	151.0	258.92
30	38	15.203	44.578	−23.127	−47.065	101.48	91.702
30	40	22.625	66.34	6.2525	12.724	151.02	258.97

Reflections

To show that the closed-form solution, MLF, can indeed replace the time-honoured iterative load flow procedure, we worked on some practical systems. Even the ill-conditioned systems noted for their obstinate non-convergent behaviour with iterative methods posed no problem with MLF!

References

1. http://www.ee.washington.edu/research/pstca/pf30/ieee30cdf.txt
2. S. Iwamoto, Y. Tamura, A load flow computational method for ill conditioned power systems. IEEE Trans. PAS **100**(4), 1736–1743 (1981)
3. S.C. Tripathy, G.D. Prasad, O.P. Malik, G.S. Hope, Load flow solutions for ill-conditioned power systems by a Newton-like method. IEEE Trans. PAS **101**(10), 3648–3657 (1982)
4. P.R. Bijwe, S.M. Kelapure, Non-divergent Fast Power Flow methods. IEEE Trans. Power Syst. **18**(2), 633–638 (2003)

Chapter 6
Outage Analysis

My nonsense is conjugate to yours....The physicist who finally succeeds in adding these two nonsenses will gain the truth!.
Wolfgang Pauli to Niels Bohr

Maintaining voltages and line flows within prescribed limits is an important security function in power systems. One of the prerequisites for this is to identify overloaded lines and then initiate preventive control action. Security assessment for major outages (N-k contingency) is required. Conducting a full load flow is fraught with issues of non-convergence and also computation time. Distribution factor approximations are employed to reduce computation time but loss in accuracy can be significant. Modular analysis makes it simple to determine the effect of outages with good accuracy.

6.1 Line Outage

Analysis of line outage with Modular Analysis is a straightforward procedure and can be stated in the form of algorithm as follows.

1. Compute bus impedance matrix and the power flows in the original network using Eqs. (4.21) and (4.22).
2. Construct new bus impedance matrix for the *outaged* network.
3. Compute new line-flows using Eqs. (4.21) and (4.22).

Remark It is to be noted that, if the line outage causes drastic changes in line flows, distribution factors would give erroneous results. With MLF there is no loss of accuracy.

M.V. Hariharan et al., *Modular Load Flow for Restructured Power Systems*, Lecture Notes in Electrical Engineering 374, DOI 10.1007/978-981-10-0497-1_6

6.1.1 Example

Power system example is shown in Fig. 6.1. Base MVA = 100.

$$Z^{base} = \begin{bmatrix} 0.6520 + 0.0104i & 0.6278 - 0.0298i & 0.6159 - 0.0520i & 0.6144 - 0.0560i & 0.6076 - 0.0654i \\ 0.6278 - 0.0298i & 0.6209 - 0.0190i & 0.6050 - 0.0536i & 0.6044 - 0.0551i & 0.6003 - 0.0577i \\ 0.6159 - 0.0520i & 0.6050 - 0.0536i & 0.6112 - 0.0246i & 0.6063 - 0.0387i & 0.5897 - 0.0745i \\ 0.6144 - 0.0560i & 0.6044 - 0.0551i & 0.6063 - 0.0387i & 0.6104 - 0.0265i & 0.5908 - 0.0715i \\ 0.6076 - 0.0654i & 0.6003 - 0.0577i & 0.5897 - 0.0745i & 0.5908 - 0.0715i & 0.6106 - 0.0104i \end{bmatrix}$$

Illustrative calculations for obtaining flow in element # 3 (line 2–3) in pre-outage network are shown below.

1. $\frac{P_1}{R_{11}^{base}} = \frac{1.2978}{0.652} = 1.99$
2. $\left[A_1^T Z^{base}\right]_{\substack{row,3\\ col,1}} = Z_{21}^{base} = 0.6278 - j0.0298$
3. $\left[A^T Z^{base}\right]_{\substack{row,3\\ col,1}} = Z_{21}^{base} - Z_{31}^{base}$

$$= (0.6278 - j0.0298) - (0.6159 - j0.0520)$$
$$= 0.0119 + j0.0222$$

4. $y_3^* = \left(\frac{1}{0.06 + j0.18}\right)^* = 1.6667 + j5.00$

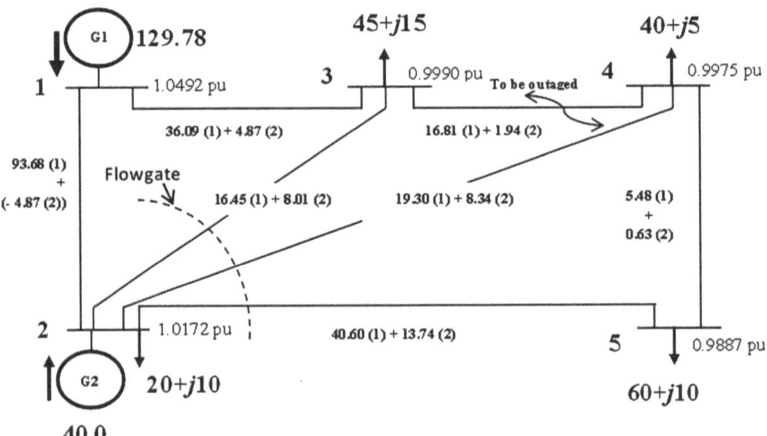

Fig. 6.1 Pre-outage network

These values are used for substitution in step 5.

5. $$p_{31f}^{base} = \mathrm{Re}\left\{\frac{P_1}{R_{11}}\left[A_1^T Z^{base}\right]_{\substack{row,3\\col,1}}\left[Z^{base*}A\right]_{\substack{row,1\\col,3}} y_3^*\right\}$$
$$= \mathrm{Re}\{1.99 * (0.6278 - j0.0298)(0.0119 - j0.0222) * (1.6667 + j5.00)\}$$
$$= 0.1645 \text{ pu}$$

Similarly for contributions from generator 2,

$$p_{32f}^{base} = 0.0801 \text{ pu}$$

Total pre-outage power flow on line # 3 (between bus 2 and bus 3) is,

$$p_{3f}^{base} = p_{31f}^{base} + p_{32f}^{base} = 0.2446 \text{ pu}$$

Voltages are calculated from (4.24). Pre-outage flows and voltages are shown in Fig. 6.1.

Outage of Line # 6 (between bus 3 and bus 4)

Now, consider line # 6, from bus 3 to bus 4 for outage. Figure 6.2 shows the outage network. New line flows and voltages are calculated from the post-outage bus impedance matrix with exactly same steps as those for the pre-outage network.

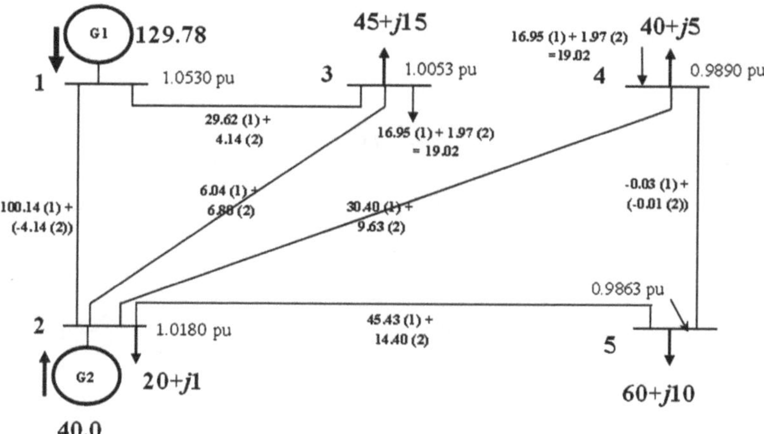

Fig. 6.2 Post-outage line flows in MW

Post-outage bus impedance matrix Z^{outage} is as shown below.

$$Z^{outage} = \left[\begin{array}{c|c|c|c|c} 0.6563 + 0.0302i & 0.6308 - 0.0133i & 0.6245 - 0.0208i & 0.6104 - 0.0551i & 0.6088 - 0.0543i \\ \hline 0.6308 - 0.0133i & 0.6235 - 0.0038i & 0.6092 - 0.0330i & 0.6040 - 0.0454i & 0.6024 - 0.0447i \\ \hline 0.6245 - 0.0208i & 0.6092 - 0.0330i & 0.6354 + 0.0467i & 0.5882 - 0.0728i & 0.5867 - 0.0720i \\ \hline 0.6104 - 0.0551i & 0.6040 - 0.0454i & 0.5882 - 0.0728i & 0.6277 + 0.0307i & 0.5973 - 0.0468i \\ \hline 0.6088 - 0.0543i & 0.6024 - 0.0447i & 0.5867 - 0.0720i & 0.5973 - 0.0468i & 0.6148 + 0.0059i \end{array}\right]$$

Flow in line # 3 (between buses 2 and 3) after outage is calculated in following steps:

1. $\frac{P_1}{R_{11}^{outage}} = \frac{1.2978}{0.6563} = 1.9774$
2. $\left[A_1^T Z^{outage}\right]_{\substack{row,3\\ col,1}} = Z_{21}^{outage} = 0.6308 - j0.0133$
3. $\left[A^T Z^{outage}\right]_{\substack{row,3\\ col,1}} = Z_{21}^{outage} - Z_{31}^{outage}$

$$= (0.6308 - j0.0133) - (0.6245 - j0.0208)$$
$$= 0.0063 + j0.0075$$

4. $y_3^* = \left(\frac{1}{0.06 + j0.18}\right)^* = 1.6667 + j5.00$
5. $p_{31f}^{outage} = \mathrm{Re}\left\{\frac{P_1}{R_{11}^{outage}}\left[A_1^T Z^{outage}\right]_{\substack{row,3\\ col,1}}\left[Z^{ouyage*}A\right]_{\substack{row,1\\ col,3}} y_3^*\right\}$

$$= \mathrm{Re}\{1.9774 * (0.6308 - j0.0133)(0.0063 - j0.0075) * (1.6667 + j5.00)\}$$
$$= 0.0604\,\mathrm{pu}$$

Similarly,

$$p_{32f}^{outage} = 0.0680\,\mathrm{pu}$$

Total power flow on line #3 (between bus 2 and bus 3), after outage is,

$$p_{3f}^{outage} = p_{31f}^{outage} + p_{32f}^{outage} = 0.1284\,\mathrm{pu}$$

Voltages are calculated from (4.25). Post-outage flows and voltages are shown in Fig. 6.2. Results are tabulated in Table 6.1.

6.2 Power Flow Fractions Versus Distribution Factors

Distribution factors are generalizations of generation shift factors [1]. In derivation of distribution factors, surplus (incremental) injected currents and powers by generators are assumed to be withdrawn at the slack bus or, some other bus. This assumption is

Table 6.1 Power flows in lines and flows (consumptions) in loads, pu

Element #	From	To	G1 contribution pre-outage	G1 contribution post-outage	G2 contribution pre-outage	G2 contribution post-outage	Total flow pre-outage	Total flow post-outage
1	1	2	0.9368	1.0014	−0.0487	−0.0414	0.8880 (0.8874)	0.9601 (0.9672)
2	1	3	0.3609	0.2963	0.0487	0.0413	0.4096 (0.4087)	0.3376 (0.3369)
3	2	3	0.1645	0.0604	0.0801	0.0680	0.2446 (0.2465)	0.1284 (0.1229)
4	2	4	0.1930	0.3041	0.0834	0.0963	0.2765 (0.2789)	0.4004 (0.4102)
5	2	5	0.4000	0.4543	0.1374	0.1440	0.5374 (0.5478)	0.5983 (0.6144)
6 *outaged*	3	4	0.1681	0.00	0.0194	0.00	0.1876 (0.1896)	0.00 (0.00)
7	4	5	0.0548	−0.0003	0.0063	−0.0001	0.0610 (0.0638)	−0.0004 (0.00039)
Load@2	2	6	0.1572	0.1574	0.0497	0.0499	0.2070 (0.20)[a]	0.2073 (0.20)[a]
Load@3	3	6	0.3422	0.3474	0.1069	0.1075	0.4491 (0.45)[a]	0.4548 (0.45)[a]
Load@4	4	6	0.3031	0.2971	0.0949	0.0941	0.3980 (0.40)[a]	0.3913 (0.40)[a]
Load@5	5	6	0.4461	0.4433	0.1405	0.1404	0.5866 (0.60)[a]	0.5837 (0.60)[a]

[a]These are data values of loads in iterative load flow

not avoidable. However, loads are impedance connections to ground which provide natural path for generator injections. The above assumption therefore appears to be superfluous. Incremental power injected at a generator bus does not *actually* go into the slack bus or into a single *other* bus; the power and associated currents return through the load impedances, as they indeed should. Load consumptions thereby increase in a manner so as to match (along with a small increase in line losses) the surplus generation. Modular Analysis takes into account this fact by representing loads as impedances.

Power flow fractions of Modular Analysis may be viewed as *differently defined* distribution factors. No assumptions, as in case of distribution factor method, are made in the new method of power fractions. Numerical results may differ due to the difference in the formulation.

6.3 Identifying Outaged Generator

We will consider a ‘short’ steady state period immediately after outage of a generator, when circuit transients have died down but inertia-related changes have not taken place. Following assumptions are made.

1. Powers injected by generators (except the outaged one) remain at their pre-outage values.
2. Circuit parameters do not change.

The algorithm can be stated as follows

a. Obtain line flows for the generator outages taken one at a time, using (4.21) and (4.22). These flow-signatures consist of line flows in Table 6.1. There are as many flow-signatures as there are generators.
b. Get measurements of the post-outage line flow from SCADA.
c. Match flows obtained in step ‘b’ with the flow-signatures obtained in ‘a’.

Agreeing flow-signature immediately identifies the outaged generator.

6.3.1 Example

For the system shown in Fig. 6.1, using Tables 6.1, 6.2 and 6.3,

Table 6.2 Flow-signature for outage of generator 1

Line	1	2	3	4	5	6	7
Flows	0.9368	0.3609	0.1645	0.1930	0.4000	0.1681	0.0548

Table 6.3 Flow-signature for outage generator 2

Line	1	2	3	4	5	6	7
Flows	−0.0487	0.0487	0.0801	0.0834	0.1374	0.0194	0.0063

Measurement pattern of line flows after a generator outage must match one of the above flow-signatures. Outaged generator can thereby be identified from line flow measurements. In practice it may not be really necessary to match all points on the flow-signature. For example, if one observes flow on line 3 dropping to 0.1645 pu MW, it is very likely that generator 1 has been outaged. If in doubt, one can confirm by checking that the flow on line 5 has indeed dropped to 0.4 pu MW, and continuing for confirmation, if required, with still another line.

6.4 Selective Computation of Line Flows

In conventional load flow, full load flow procedure must be completed even if one is interested in power flows only on a few selected lines. MLF is very helpful in such situations. Selected lines could be one or more tielines, flowgates, or even a single line of critical importance. In MLF, individual line flows are obtained independently (see Eq. (4.21)) and results are then collated to form the table of flows. We illustrate this by an example.

6.4.1 Example

Lines # 3, 4 and 5 constitute a flowgate in power system shown in Fig. 6.1. We are required to calculate only 3 line flows. We do this in the pre-outage network by calculating $p_{31f}, p_{32f}, p_{41f}, p_{42f}, p_{51f}$ and p_{32f} using (4.20)–(4.22). These values are available in Table 6.1. Thereby we obtain,

$$\begin{aligned} p_{3f} &= p_{31f} + p_{32f} = 0.1645 + 0.0801 = 0.2446\,\text{pu} \\ p_{4f} &= p_{41f} + p_{42f} = 0.1930 + 0.0834 = 0.2765\,\text{pu} \\ p_{5f} &= p_{51f} + p_{52f} = 0.40 + 0.1374 = 0.5374\,\text{pu} \\ p_{flowgate} &= p_{3f} + p_{4f} + p_{5f} = 1.0585\,\text{pu} \end{aligned}$$

Reflections

In Modular Load Flow there is no slack bus, nor is it necessary to have a base case solution that is necessary to obtain sensitivities and distribution factors to perform conventional contingency analysis. Exact solution is available via *MLF. Computation in MLF is extremely fast* (6 *to* 8 *times the conventional Newton Raphson) making it suitable for online application. No additional data is required. Only a small set of instructions added to the existing packages would do the trick!*

Reference

1. A.J. Wood, B.F. Wollenberg, *Power Generation, Operation And Control*, 2nd edn. (John Wiley and Sons, New York, 1996)

Chapter 7
Voltage Behaviour

Get the Physics right, rest is Mathematics ...
Rudolf Kalman

Poor reactive power support leads to voltage reduction and may result in voltage collapse. Exhaustive literature exists on voltage stability studies. Almost all studies specify voltage at the source end. However under stressed conditions, system may not have enough reactive power to support voltages. Even the generator voltages might get depressed due to *loading* effect. It then behaves as a power source. Voltage behavior of systems with power source has not been studied so far. This chapter is devoted to such a study. As we shall find out in this chapter, voltage behaviour with constant power source turns out to be at surprising variance with those discussed in literature hitherto.

PV and QV curves are normally obtained with constant voltage specified at the source end [1]. These curves have been extensively used for study of voltage collapse phenomena. The curves are valid under normal conditions when generators have enough reserves of reactive powers and can maintain sending end voltage. The results however may be significantly erroneous under extreme loading conditions with generators hitting their capability limits and acting as constant power source. Such conditions can trigger critical outages causing cascading failures and blackouts. In the pre-cascading regime, the load consumptions cannot *factually* increase since generators having hit limits cannot deliver any more power. Loads receive only as much power as the generators can provide. Performance of the system under such conditions is dominated by *power constraints*. Right question therefore to ask is, "what voltages will appear in the system for a given set of real and reactive *injected* powers?" It may be recalled that during the iterative process of the conventional or continuation load flow, one needs to check 'Q' for PV buses, switch them to PQ-buses if limits are violated, and again switch back to PV if they revert within limits, or, alternatively, lower the specified voltage in steps until we get the 'Q' within limits. Entire procedure is seen to be driven by arbitrary voltage and Q-limit specifications. Precise information about these is usually not available to the operator and results may significantly deviate from operational reality. Our calculations start with the given values of injected powers and computes behavioral changes in system voltages.

M.V. Hariharan et al., *Modular Load Flow for Restructured Power Systems*,
Lecture Notes in Electrical Engineering 374, DOI 10.1007/978-981-10-0497-1_7

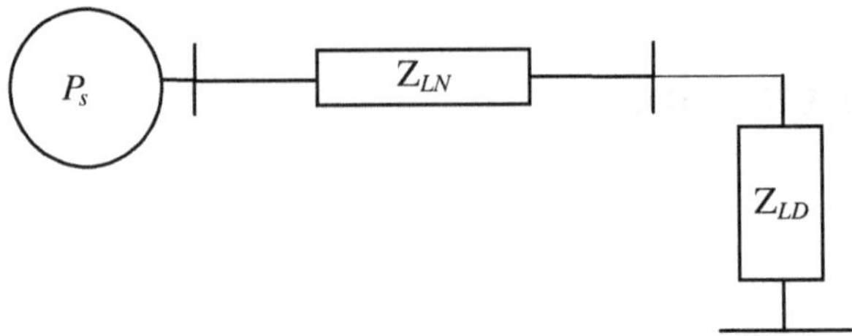

Fig. 7.1 Basic system with power source

7.1 Problem Formulation

Figure 7.1 shows a source connected to a line of impedance $Z_{LN}\angle\theta$ and a load of impedance $Z_{LD}\angle\phi$. Two cases will be discussed. In the first case, the source is a conventional constant voltage source. In the second case, which illustrates the new approach, the source is an ideal power source which injects a constant power P_s into the network. Simple derivations below show expressions for currents, voltages and powers at the receiving end for the two cases [1].

Kundur [1] (voltage source)	Modular load flow (power source)
$I_s = \frac{V_s}{Z_{LN}\sqrt{1+\left(\frac{Z_{LD}}{Z_{LN}}\right)^2+2\left(\frac{Z_{LD}}{Z_{LN}}\right)\cos(\theta-\phi)}}$ (7.1a)	$I_s = \sqrt{\frac{P_s}{Z_{LN}\cos\theta+Z_{LD}\cos\phi}}$ (7.1b) $= \sqrt{\frac{\left(\frac{P_s}{Z_{LN}}\right)}{\cos\theta+\frac{Z_{LD}}{Z_{LN}}\cos\phi}}$
$V_R = \frac{V_s}{\sqrt{1+\left(\frac{Z_{LD}}{Z_{LN}}\right)^2+2\left(\frac{Z_{LD}}{Z_{LN}}\right)\cos(\theta-\phi)}}\frac{Z_{LD}}{Z_{LN}}$ (7.2a)	$V_R = I_sZ_{LD} = \sqrt{\frac{P_sZ_{LD}}{\cos\phi+\frac{Z_{LN}}{Z_{LD}}\cos\theta}}$ (7.2b)
$P_R = \frac{Z_{LD}}{1+\left(\frac{Z_{LD}}{Z_{LN}}\right)^2+2\left(\frac{Z_{LD}}{Z_{LN}}\right)\cos(\theta-\phi)}\left(\frac{V_s}{Z_{LN}}\right)^2\cos(\phi)$ (7.3a)	$P_R = (I_s^2Z_{LD}\cos\phi) = \frac{P_s}{1+\left(\frac{Z_{LN}}{Z_{LD}}\right)\frac{\cos\theta}{\cos\phi}}$ (7.3b)

Expressions (7.2b)–(7.3b) are obtained with power source and are different from those with constant voltage source, i.e., (7.2a) and (7.3a). Alternative expressions for (7.2b) and (7.3b) can be derived with modular approach.

7.2 Modular Approach

Let,

$$\begin{aligned} Z_{LN} &= r_1 + jx_1 \\ Z_{LD} &= r_2 + jx_2 \end{aligned} \tag{7.4}$$

Using,

$$\begin{aligned} R &= r_1 + r_2 \\ X &= x_1 + x_2 \end{aligned} \tag{7.5}$$

$$I = \sqrt{\frac{P_s}{R}} \tag{7.6}$$

Therefore magnitude of voltage and power at receiving end is,

$$|V_r| = |IZ_{LD}| = \sqrt{\frac{P_s}{R}}|(Z_{LD})| \tag{7.7}$$

$$P_r = \left(\frac{P_s}{R}\right) r_2 \tag{7.8}$$

Expressions (7.6), (7.7) and (7.8) are equivalent to (7.1b), (7.2b) and (7.3b), respectively. Current, voltage and load-powers with constant power source are obtained using (7.6), (7.7) and (7.8) and are shown in Fig. 7.2b. With constant voltage source these are obtained from (7.1a), (7.2a) and (7.3a) and shown in Fig. 7.2a. PV curves with the two sources are shown in Fig. 7.3a and 7.3b.

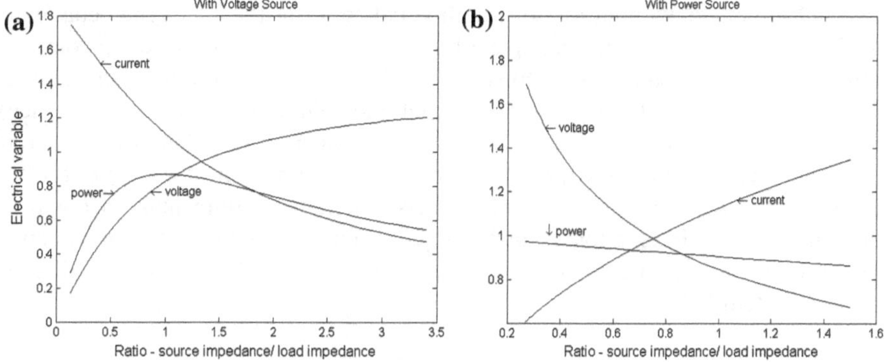

Fig. 7.2 Variation of electrical variables with 'Voltage' and 'Power' as sources

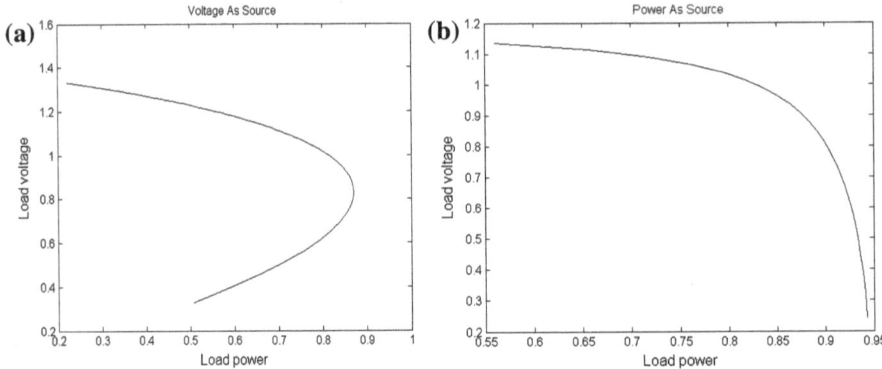

Fig. 7.3 PV curves with 'Voltage' and 'Power' as sources

An important difference in Fig. 7.2a, and 7.2b is the manner in which voltages and currents vary in power-sourced and voltage-sourced circuits. Nature of their variations is reversed! With voltage source, the *current* decreases monotonically. With power-source, it is the *voltage* that decreases monotonically. This reversal in nature of voltage variation strikingly affects the PV curves; with voltage source we get a *nose* in the curve, with power source there is no *nose* but a *knee* (Fig. 7.3b). Nose point has been been used to indicate phenomenon of bifurcation which has been subject of intensive research [2].

We observe from Fig. 7.3a that mapping between power-to-voltage is one-to-many (one-to-two in this case) with constant voltage source. This is the reason why usual power flow procedure results in multiple solutions, and, why near the nose of the curve, non-convergence becomes a serious issue. With power-sourced circuits the mapping from power-to-voltage is one-to-one always (Fig. 7.3b), hence unique solution results. The so-called *nose* (voltage instability) does not appear in PV curves with power-sources. It is the *knee* that marks departure (but not instability) in voltage behaviour of the system. Practical behaviour of voltage collapse described in [1, Chap. 14] shows first a gradual and then a rapid fall in voltage without loss of stability and confirms the knee type behaviour. MLF illustrated for simple example of Fig. 7.1 is easily extended to complex networks. Iterative load flow, conventional NR method or Continuation method, may play truant (singularity of jacobian, multiple solutions and non-convergence) under low voltage situations [3, 4]. Condition number of jacobian is often a criterion whether or not to expect convergence. Continuation load flow further incorporates a *lambda* parameter which makes it rather judgmental. In contrast, solution with MLF is explicit and unique even for multi-generator systems. Condition number is of no relevance in MLF calculations. We discuss an ill conditioned system below.

7.2.1 Example: Ill-Conditioned System

The system shown in Fig. 7.4 has been extensively researched for its obdurate non-convergence; see for example [5, 6]. It has high and low voltage buses located close to each other. The line parameters are obtained from off-diagonal terms of bus admittance matrix given in [6] by assuming transformer ratios, if any, as unity. Half line charging of 0.0001 pu is added to line data. The program adds a small resistance of 0.0001 to net shunt admittance at a bus if it does not have a real part. This is done to enable use of expression (4.25) for calculation of bus voltages. Line data, generator data and load data are given in Tables 7.1, 7.2 and 7.3. Difference between these and the original data in [6] are very small and validity of conclusion from results of this system is not affected. Base MVA is 100 MVA.

NR method with flat start fails to converge for this system. It is reported to be a working system. A solution therefore must exist. Rectangular version of ILF converges [6]. Possibility of decoupling of load flow equations is however lost in the rectangular version. Tamura and Iwamoto [5] proposed a new set of equations in polar form in which a parameter μ is introduced which is tuned iteratively to obtain convergence. MLF is easily conducted on this system. Decoupling becomes a non-issue and issues of non-convergence or multiple solutions simply do not arise since the solution is non-iterative. Load voltages from MLF are found to be $V_3 = 0.7221$, $V_5 = 0.8013$, $V_6 = 1.0487$, $V_9 = 0.7042$, and $V_{11} = 0.9210$ (Table 7.4). The voltage profile is not a usual one.

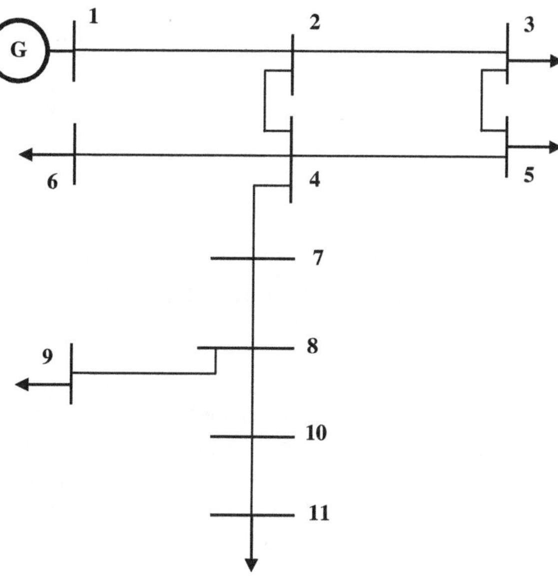

Fig. 7.4 Ill-conditioned 11-bus example

Table 7.1 Line data, pu

Line #	From	To	R	X	B
1	1	2	0.0	0.0707	0.0001
2	2	3	0.0	0.1540	0.0001
3	2	4	0.0377	0.0413	0.0001
4	3	5	0.1228	0.1803	0.0001
5	4	5	0.0	0.4593	0.0001
6	4	6	0.0	0.0176	0.0001
7	4	7	0.6114	0.8118	0.0001
8	7	8	0.1621	0.2167	0.0001
9	8	9	0.0718	0.7180	0.0001
10	8	10	0.4098	0.5600	0.0001
11	10	11	0.0246	0.2646	0.0001

Table 7.2 Generator data

Gen no.	At bus	MW + *j*MVAR
1	1	60 + *j*27.6

Table 7.3 Load data

Load no.	At bus	MW	MVAR
1	3	12.8	6.2
2	5	16.5	8.0
3	6	9.0	6.8
4	9	2.6	0.9
5	11	15.8	5.7

Table 7.4 Voltage variation at buses with load (equivalent impedance) change at bus 5

		Multiple of nominal load at bus 11								
Ld#	Voltage @bus	0.85	0.875	0.90	0.925	0.95	0.975	**1.00**	1.025	1.05
1	3	0.7987	0.7846	0.7711	0.7581	0.7456	0.7336	**0.7221**	0.7110	0.7004
2	5	0.8970	0.8794	0.8624	0.8462	0.8306	0.8157	**0.8013**	0.7875	0.7742
3	6	1.1760	1.1525	1.1300	1.1084	1.0877	1.0678	**1.0487**	1.0303	1.0126
4	9	0.8081	0.7889	0.7705	0.7529	0.7360	0.7198	**0.7042**	0.6892	0.6748
5	11	1.0613	1.0354	1.0105	0.9867	0.9639	0.9421	**0.9210**	0.9008	0.8814

7.3 Variation of Voltage

Several loading conditions have to be used for obtaining variation in voltages at different buses. Use of ILF is seriously handicapped by non-convergence and becomes problematic for higher loads. The MLF proposed here makes this study rather straightforward. We show how voltages at different buses vary with load variation at bus 11. Calculations are *one-shot*, i.e., they need to be done only once (no iteration!) for each value of load. The PV curves can then be plotted for voltages of each bus against *load* at bus 11. This is shown graphically in Fig. 7.5 which has been obtained from computations on full system and not a source-line-load type equivalent of Fig. 7.1. It may be noted that convergence of even normal load flow is difficult for this system with iterative load flows, but MLF succeeds in obtaining the entire PV curve. In Fig. 7.5, abscissa represents variation in load at bus 11. For MLF, PQ load at each step is converted into impedance form.

7.4 Voltage Dependency of Loads

Voltage dependency of loads is represented by second degree polynomials in voltage variable, $P = a1 * V^2 + b1 * V + c1$; $Q = a2 * V^2 + b2 * V + c2$ [7]. In MLF, this polynomial turns out to be, only the square term of the polynomial, i.e., for $R - X$ representation, $G + jB = 1/(R + jX)$. Then, $P = G * V^2$, and, $Q = B * V^2$. The linear and constant terms do not appear.

$$P = P_0\left(\frac{V}{V_0}\right)^2 \tag{7.9}$$

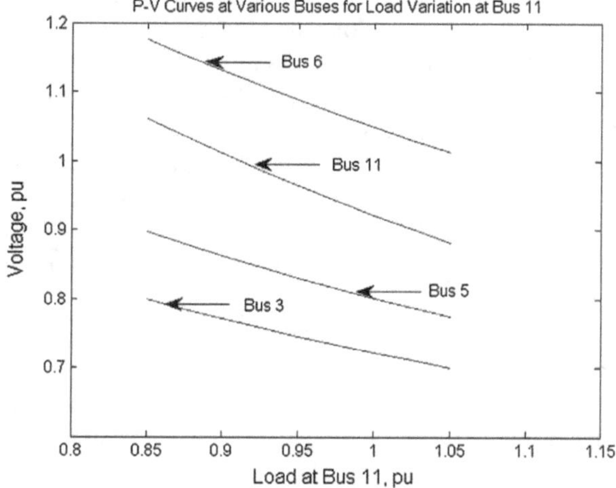

Fig. 7.5 'Load@Bus-11' versus voltage variation at various nodes

and

$$Q = Q_0 \left(\frac{V}{V_0} \right)^2 \tag{7.10}$$

And we have,

$$\frac{Q}{P} = \frac{X}{R} \tag{7.11}$$

We assume frequency to remain constant hence X does not change with loading. It is only the ratio $\frac{X}{R}$ that changes due to change in resistance that determines power consumed by the device (for example, in induction motors' equivalent circuits).

To deal with practical models with additional linear and constant components of loads, one may assume two extreme possible values of voltages for calculating load impedances and conduct *two* runs of MLF. The resulting voltage profiles then would provide the band within which the actual voltage must lie.

7.5 PQ-reserve and PQ-deficit Condition

In practical operation, two types of operating conditions exist; (i) with real and reactive power *reserve* available, and, (ii) with real and reactive power (*P, or, Q*) in *deficit*. In the above example, conditions are considered to be *P and Q*—deficit conditions. When *PQ* reserves are available, voltage at the generator bus can be maintained constant and the specified load powers can be met. Just prior to a blackout, all generator reserves are generally exhausted. *Before* the onset of this condition, constant voltage source methods apply. Consider system in Fig. 5.1 with generator having Q-capability to maintain voltage of 1.3 pu. With this voltage, source is always able to supply load specified at bus 5. With voltage of 1.45 pu at the source, power in load impedance becomes more than 1.0 pu at the point of intersection, which the source cannot supply. Thereafter generator power is pegged at 1.0 pu and voltage declines along the starred *constant power source* PV curve (Fig. 7.6). Depending on the capability limits, this switchover may or may not occur in practice. With sending end voltage (lesser reactive power limit) of 1.30 pu, and real power limit of 1.0 pu, switchover does not occur and voltage collapse ensues via the nose point of the constant-voltage-source PV curve. With source voltage of 1.45 pu voltage collapse will traverse the starred curve beyond the point of intersection.

With or without the switchover, voltage declines faster than before, leading eventually to voltage collapse. In case of power source, load voltage falls *rapidly* beyond some critical point but continues to be stable even at low voltages. In case of constant voltage source the voltage instability sets in at nose point. Practical scenarios described in (Chap. 14, [1]) show that voltages in system continue to

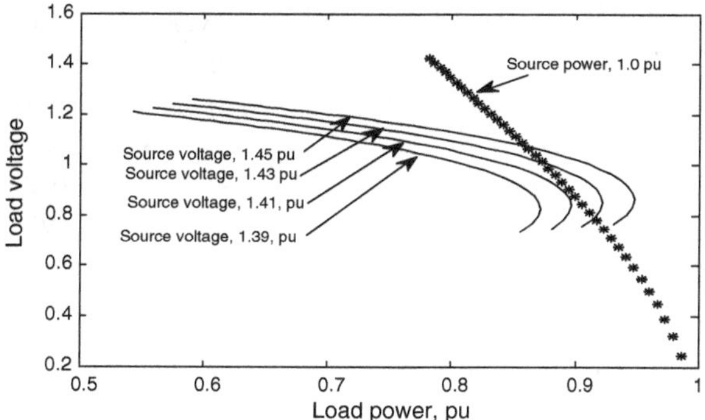

Fig. 7.6 Effect of source type on P-V curves

stably exist for all loading points even in fast declining region vindicating our premise that generators act as power source for extreme loads. Voltage levels however indicate *vulnerability*, as small changes in load would result in disproportionately large changes in voltage.

7.5.1 Example: The Switchover

Figure 7.7 shows a 5 bus example [8]. We reproduce data in Tables 7.5 and 7.6 for ready reference. Base: 10 MVA.

Load impedance is varied at bus 5. To obtain starred curve shown in Fig. 7.6, source power is maintained at 1.0 pu and MLF is conducted. Impedance

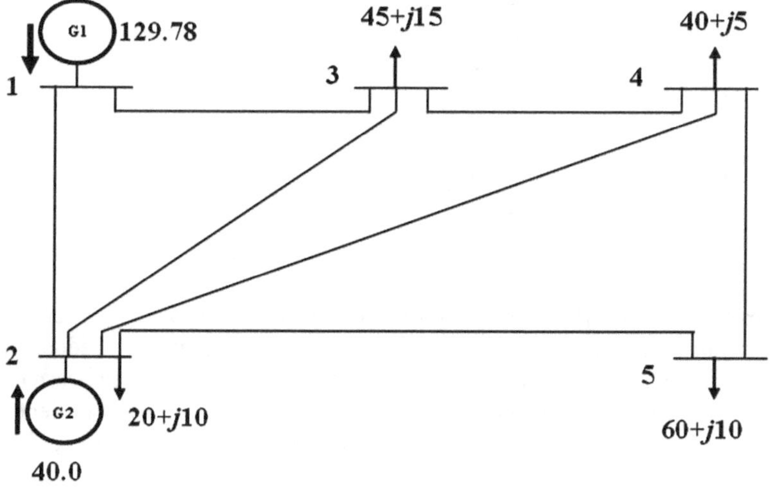

Fig. 7.7 A 5-bus example

Table 7.5 Bus data, pu

Node	|Voltage|	PG	QG	PL	QL
1	1.06	1.2978	0.2826	0.0	0.00
2	1.03	0.40	−0.0638	0.20	0.10
3	1.0118	0.0	0.0	0.45	0.15
4	1.0102	0.0	0.0	0.40	0.05
5	1.0014	0.0	0.0	0.60	0.10

Table 7.6 Line data, pu

No.	R	X	½ B charging
1–2	0.02	0.06	0.03
1–3	0.08	0.24	0.025
2–3	0.06	0.18	0.02
2–4	0.06	0.18	0.02
2–5	0.04	0.12	0.015
3–4	0.01	0.03	0.010
4–5	0.08	0.24	0.025

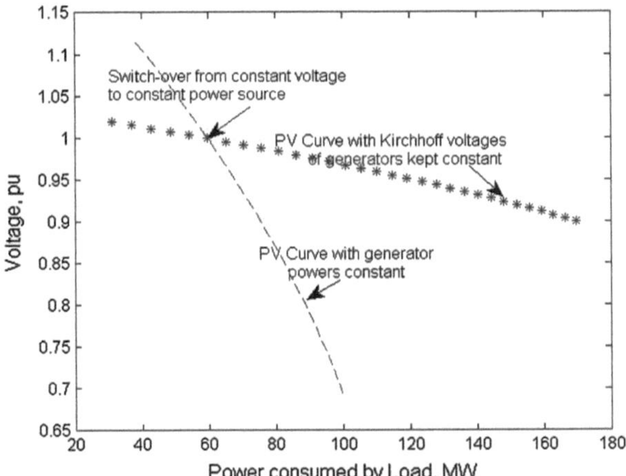

Fig. 7.8 Switchover of generators from constant voltage to constant power source

corresponding to 60 + j10 MW gives 1.0 pu voltage at load bus (#5). Sending end voltage required for this is 1.45 pu. If impedance (real component) is reduced below this (to increase real power) the voltage falls. It would fall along the starred curve with generator working as source of 1.0 pu power. Algorithm given in Sect. 4.7 is used. Region near the intersection point is zoomed in and shown in Fig. 7.8.

7.6 Conclusion

Study of voltage behaviour during overstressed conditions of power systems is of vital importance, especially in deregulated environment where maintaining voltage (reactive power management) has become a serious issue; Gencos would not produce reactive powers on their own as there is no incentive to do so. Traditional power flow procedures currently used by operators cannot specify PV bus voltages with good reliability. There is a genuine threat of non-convergence of loadflow under low voltage conditions. MLF works for all cases. That the phase angles do not appear in this analysis is a great computational relief as all trigonometric functions are avoided.

Reflections

Analysis of electrical circuits with power source sprang a surprise on us. We found, and analytically proved, that with power-constrained circuits there is no 'nose' in the PV curves. We find ourselves at variance with several researchers exploring bifurcation phenomena occurring at this 'nose' point. In our scheme a rapid fall in voltage does occur after certain loading, but it is not the kind of decline that a 'nose' would predict! All points in the declining region are stable operating points even when the decline is extremely fast.

References

1. P. Kundur, N.J. Balu, M.G. Lauby, *Power System Stability and Control* (McGraw-Hill Education, 1994)
2. I. Dobson, T. Van Cutsem, C. Vournas, C.L. DeMarco, M. Venkatasubramanian, T. Overbye, C. A. Canizerous, *Voltage Stability Assessment: Concepts, Practices and Tools*, Ch 2. IEEE Power Engineering Society, Power System Stability Subcommittee Special Publication, Aug 2002
3. T.J. Overbye, A power flow measure for unsolvable cases. IEEE Trans. Power Syst. **9**(3), 1359–1365 (1994)
4. T.J. Overbye, R.P. Klump, Effective calculation of power system low voltage solutions. IEEE Trans. Power Syst. **11**(1), 75–82 (1996)
5. S. Iwamoto, Y. Tamura, A load flow method for ill-conditioned power systems. IEEE Trans. PAS **100**(4), April 1981
6. S.C. Tripathy, G.D. Prasad, O.P. Malik, G.S. Hope, Load flow solutions for ill-conditioned power systems by a Newton-like method. IEEE Trans. PAS, PAS **101**(10), 3648–3657, Oct 1982
7. J.V. Milanović, K. Yamashita, S. Martinez Villanueva, S.Z. Djokić, L.M. Korunović, International industry practice on power system load modeling. IEEE Trans. Power Syst. **28**(3), 3038–3046, August 2013
8. G.W. Stagg, A.H. El-Abiad, *Computer Methods in Power System Analysis*, 1st edn. (McGraw-Hill Inc. US, 1968)

Chapter 8
Optimization

Everything should be made as simple as possible, but not simpler…

Albert Einstein

Optimization implies minimization or maximization of some objective function. Powers, line losses, line flows etc. are nonlinear functions of phasor variables. Optimization therefore involves solution of simultaneous partial differential equations. System constraints must be satisfied by the solution variables. A Lagrangian is usually constructed and solution obtained by solving partial differential equations [1]. In Modular approach, line losses, line flows, squares of voltages, are shown to be linear functions of injected powers. Most optimization problems therefore turn out to be in linear programming (LP) format. In this chapter we discuss these new formulations.

8.1 Optimal Use of Injected Powers

We pose the optimization problem as: *With sum of injected powers held constant, what allocation of injections will result in maximum consumption in loads—in the impedances representing them?*

Consider a system with two generators, with identical fuel cost curves and connected at the two ends of a transmission line. Only one load is connected at the bus of generator 1. It is obvious that maximum possible power to the load must be supplied by generator 1, since there is no transmission loss involved. Remaining requirement, *if any*, may be supplied by generator 2. This ensures that the power supplied by two generators is maximally consumed in the load impedance. The formulation maximizes use of specified injections. Optimal use can also be interpreted as *minimization of transmission loss*. For simple examples, such as the above two generator configuration, the problem can be solved merely by observation. Problem with multi-generator systems having different fuel cost curves is discussed below.

M.V. Hariharan et al., *Modular Load Flow for Restructured Power Systems*, Lecture Notes in Electrical Engineering 374, DOI 10.1007/978-981-10-0497-1_8

8.2 Multi-generator Case

Using (4.14), power consumptions in loads are given by (8.1) below. Load element '*e*' is denoted by extended symbol *Lde* where *Ld*1 means load 1; *Ld*2 means load 2, and so on.

$$\begin{bmatrix} p_{Ld1} \\ p_{Ld2} \\ \cdot \\ \cdot \\ p_{Ldnd} \end{bmatrix} = \begin{bmatrix} \varepsilon_{Ld1,1} & \varepsilon_{Ld1,2} & \cdots & \varepsilon_{Ld1,ng} \\ \varepsilon_{Ld2,1} & \varepsilon_{Ld2,2} & \cdot & \varepsilon_{Ld2,ng} \\ \cdot & \cdot & \cdot & \cdot \\ \cdot & \cdot & \cdot & \cdot \\ \varepsilon_{Ldnd,1} & \cdot & \cdot & \varepsilon_{Ldnd,ng} \end{bmatrix} \begin{bmatrix} P_1 \\ P_2 \\ \cdot \\ P_{ng} \end{bmatrix} \tag{8.1}$$

Total power consumption in loads is,

$$P_{Total-in-loads} = \sum_{e=1}^{nd} p_{Lde} = \sum_{i=1}^{ng} \sum_{e=Ld1}^{Ldnd} \varepsilon_{ei} P_i \tag{8.2}$$

The above equation can be reorganized as,

$$P_{Total-in-loads} = a_1 P_1 + a_2 P_2 + \cdots + \cdots + a_{ng} P_{ng} \tag{8.3}$$

where,

$$a_i = \left(\sum_{e=Ld1}^{Ldnd} \varepsilon_{ei} \right) \tag{8.4}$$

Optimal use is then mathematically stated as,

$$\max\{a_1 P_1 + a_2 P_2 + \cdots + \cdots + a_{ng} P_{ng}\} \tag{8.5}$$

subject to,

1. $P_1 + P_2 + \cdots + P_{ng} = K(a\,constant)$ (8.6)

2. $0 \leq P_i \leq P_i^{\max}, for\, i = 1, 2, \ldots, ng$ (8.7)

Feasibility of solution is assumed. Coefficients (a's) in (8.3) can be evaluated from parameters and topology of the network using (8.4). Due to constraint (8.6), with unity coefficient for all variables, the linear programming solution gets further simplified and boils down to simply ordering of generator allocations in decreasing order of their influence on the objective function. We illustrate this with an example.

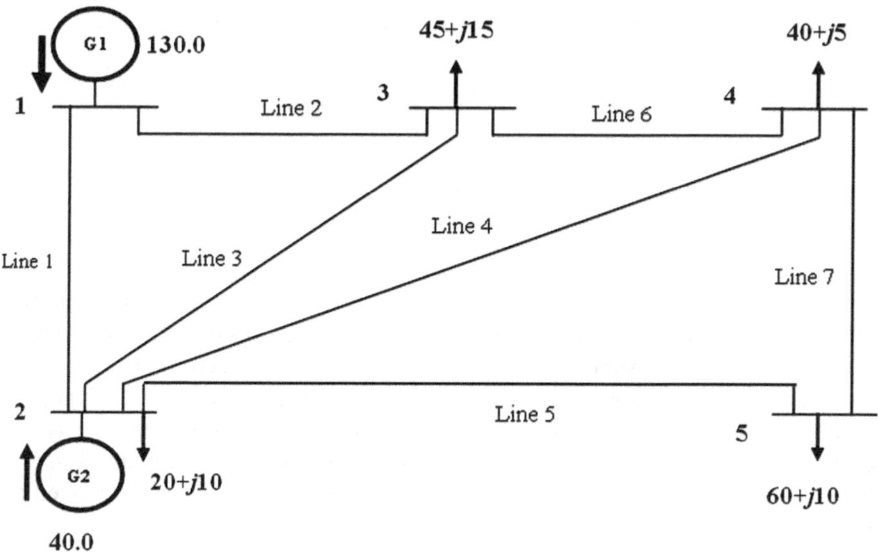

Fig. 8.1 A 5-Bus system

8.2.1 Example

Consider example from [2] shown in Fig. 8.1. We are required to find generator injections (their sum remaining constant) that will result in maximum consumption in loads (or, minimum transmission loss). Let an additional constraint $P_2^{\max} = 75.0$ be enforced. Injections in steady state are 130 MW (generator 1) and 40 MW (generator 2). Power fractions for load elements, obtained from the network data (see (3.2)–(3.4)). These are,

$$\begin{aligned}&\varepsilon_{Ld1,1} = 0.1213,\ \varepsilon_{Ld2,1} = 0.2636,\ \varepsilon_{Ld3,1} = 0.2335,\ \varepsilon_{Ld4,1} = 0.3440\\&\varepsilon_{Ld1,2} = 0.1249,\ \varepsilon_{Ld2,2} = 0.2667,\ \varepsilon_{Ld3,2} = 0.2371,\ \varepsilon_{Ld4,2} = 0.3525\end{aligned} \tag{8.8}$$

Power consumption in loads are given by (8.1), i.e.,

$$\begin{bmatrix} p_{Ld1} \\ p_{Ld2} \\ p_{Ld3} \\ p_{Ld4} \end{bmatrix} = \begin{bmatrix} 0.1213 & 0.1249 \\ 0.2636 & 0.2667 \\ 0.2335 & 0.2371 \\ 0.3440 & 0.3525 \end{bmatrix} \begin{bmatrix} P_1 \\ P_2 \end{bmatrix} \tag{8.9}$$

Coefficients a's in (8.4) are obtained from sums of columns of (8.9) as,

$$a_1 = \sum_{e=Ld1}^{Ld4} \varepsilon_{e1} = 0.9623 \quad (8.10)$$

$$a_2 = \sum_{e=Ld1}^{Ld4} \varepsilon_{e2} = 0.9812 \quad (8.11)$$

Coefficients (8.10) and (8.11) indicate that a higher fraction of power of generator 2 is consumed in loads as compared to that of generator 1. Therefore maximum capacity of generator 2 (75 MW) should be utilized *first*. Remaining consumption, if any, should be supplied by generator 1. Exact total consumption required to calculate the *remaining* requirement is however not known apriori because we do not know the transmission losses. Neglecting transmission losses in the two cases, we assign (170 − 75 = 95) MW to generator 1. Results obtained are given below.

Case 1 [2]: G1 = 130 MW and G2 = 40 MW: Load consumptions are: 125.10 MW supplied by generator 1 and, 39.249 MW supplied by generator 2; Total = 164.349 MW.
Case 2: (Optimal use) G2 = 75 MW and G1 = 95 MW Load consumptions are 91.42 from generator 1 and 73.592 from generator 2; Total = 165.012 MW.

To confirm that we are indeed at optimum, let us shift 5 MW in case 2, from generator 2 to generator 1, and assign 100 and 70 MW to G1 and G2, respectively. Then the load consumptions are found to be, 96.23 MW from generator 1 and 68.68 MW from generator 2; Total = 164.917 MW which is less than the optimum.

Analysis in this section can be readily extended to *delivery cost minimization* by suitably modifying (8.1).

8.3 Load Voltage Optimization

The problem is stated as: *what generator allocations will result in minimizing the voltage deviations in bus voltages?* Using (4.14), (4.25) and the fact that, $\frac{q_{ei}}{p_{ei}} = \frac{x_e}{r_e}$ *for all i*, expression for bus voltage of load e is given by,

$$v_{Lde}^2 = p_{Lde} r_{Lde}\left(1 + \frac{x_{Lde}^2}{r_{Lde}^2}\right) \quad (8.12)$$

$$= r_{Lde}\left(\sum_{g=1}^{ng} p_{Lde,g}\right)\left(1 + \frac{x_{Lde}^2}{r_{Lde}^2}\right) \quad (8.13)$$

$$= r_{Lde}\left(\sum_{g=1}^{ng}\varepsilon_{Lde,g}P_g\right)\left(1+\frac{x_{Lde}^2}{r_{Lde}^2}\right) \tag{8.14}$$

$$v_{Lde}^2 = r_{Lde}\left(1+\frac{x_{Lde}^2}{r_{Lde}^2}\right)\left(\varepsilon_{Lde,1}P_1+\varepsilon_{Lde,2}P_2+\cdots+\varepsilon_{Lde,ng}P_{ng}\right) \tag{8.15}$$

Or,

$$v_{Lde}^2 = \sum_{g=1}^{ng} b_{eg}P_g = b_{e1}P_1 + b_{e2}P_2 + \cdots + b_{e,ng}P_{ng} \tag{8.16}$$

where,

$$b_{eg} = r_{Lde}\left(1+\frac{x_{Lde}^2}{r_{Lde}^2}\right)\varepsilon_{Lde,g} \tag{8.17}$$

Equation (8.16) can be written in matrix form as,

$$\begin{bmatrix} v_{Ld1}^2 \\ v_{Ld2}^2 \\ \cdot \\ \cdot \\ v_{Ld,nd}^2 \end{bmatrix} = \begin{bmatrix} b_{11} & b_{12} & \cdots & b_{1,ng} \\ b_{21} & b_{22} & \cdots & b_{2,ng} \\ \cdot & \cdot & \cdot & \cdot \\ \cdot & \cdot & \cdot & \cdot \\ b_{nd,1} & & & b_{nd,ng} \end{bmatrix} \begin{bmatrix} P_1 \\ P_2 \\ \cdot \\ P_{ng} \end{bmatrix} = BP \tag{8.18}$$

Subscript nd denotes number of loads. Coefficient matrix on the RHS is a rectangular matrix. Let, v_{Lde}^2 be denoted by optimization variable x_e. Then the problem of load voltage optimization can be stated as:

$$\text{Min}\{(x_1-1)+(x_2-1)+\cdots+(x_{nd}-1)\} \tag{8.19}$$

subject to,

$$\sum_{i=1}^{ng} P_i = K \tag{8.20}$$

that is,

$$\left(\sum_{j=nd,1}^{nd} b_{j1}\right)P_1 + \left(\sum_{j=nd,1}^{nd} b_{j2}\right)P_2 + \cdots, \left(\sum_{j=nd,1}^{nd} b_{j,ng}\right)P_{ng} = K \tag{8.21}$$

Or,

$$c_1P_1 + c_2P_2 + \ldots, c_{ng}P_{ng} = K \tag{8.22}$$

Inequality constraint due to upper limit, 100 MW of generator 1, is considered in formulation. Again, it is an LP problem.

8.3.1 Example

For the system shown in Fig. 8.1, we are required to optimize voltages in the quadratic sense as mentioned above. Using (8.8), (8.17) to obtain the coefficient matrix B (as defined in (8.18)),

$$B = \begin{bmatrix} 0.6065 & 0.6247 \\ 0.5857 & 0.5927 \\ 0.5837 & 0.5927 \\ 0.5733 & 0.5875 \end{bmatrix} \tag{8.23}$$

The c coefficients in the objective function (8.19) are given by,

$$\begin{aligned} c_1 &= \textit{sum of column } 1 = 2.3492 \\ c_2 &= \textit{sum of column } 2 = 2.3976 \end{aligned} \tag{8.24}$$

From observation of (8.24) we see that maximum power should be assigned to the generator with minimum c. This means that 100 MW (maximum capacity) should be first assigned to generator 1. Remaining requirement (170 − 100) is then assigned to generator 2. The solution to above optimization problem therefore is,

$$P_1 = 100\ \text{MW(max) and } P_2 = 70\ \text{MW} \tag{8.25}$$

Equation (8.18) in this case is,

$$\begin{bmatrix} v_{Ld1}^2 = x_1 \\ v_{Ld2}^2 = x_2 \\ v_{Ld3}^2 = x_3 \\ v_{Ld4}^2 = x_4 \end{bmatrix} = \begin{bmatrix} 0.6065 & 0.6247 \\ 0.5857 & 0.5927 \\ 0.5837 & 0.5927 \\ 0.5733 & 0.5875 \end{bmatrix} \begin{bmatrix} 100 \\ 70 \end{bmatrix} \tag{8.26}$$

Load voltages are then obtained from,

$$\begin{bmatrix} v_{Ld1} = \sqrt{x_1} \\ v_{Ld2} = \sqrt{x_2} \\ v_{Ld3} = \sqrt{x_3} \\ v_{Ld4} = \sqrt{x_4} \end{bmatrix} = \begin{bmatrix} 1.0217 \\ 1.0003 \\ 0.9993 \\ 0.9922 \end{bmatrix} \tag{8.27}$$

With the original scheduling of 130 MW to generator 1 and 40 MW to generator 2 the voltages are found to be,

$$\begin{bmatrix} v_{Ld1} \\ v_{Ld2} \\ v_{Ld3} \\ v_{Ld4} \end{bmatrix} = \begin{bmatrix} 1.019 \\ 0.9992 \\ 0.9980 \\ 0.9901 \end{bmatrix} \tag{8.28}$$

Improvement in voltages can be clearly seen in (8.27) as compared to those in (8.28).

One may find it rather odd that *voltage* improvement is being achieved by manipulating injected *real powers*. It may be noted any real power injection must be accompanied by corresponding reactive power. If that reactive power support is not available, something strange happens in the system due to the physics that governs it; for example, the frequency which has been considered constant hitherto, will change (thereby changing system reactances and hence the reactive power requirement). That is what happens when over-excitation-limiters (OELs) reach their limits.

8.4 Flow Optimization

We pose the problem of flow optimization as:

What values of generator powers will result in maximum reserve margin on transmission lines?

For the *nl* transmission lines in the system, let their maximum permissible flows be $P_{f1}^{\max}, P_{f2}^{\max}, P_{f3}^{\max}, \ldots, P_{fnl}^{\max}$. Using (4.22) line flows can be written in matrix form as,

$$\begin{bmatrix} P_{f1} \\ P_{f2} \\ \cdot \\ \cdot \\ P_{fnl} \end{bmatrix} = \begin{bmatrix} \varepsilon_{f1,1} & \varepsilon_{f1,2} & \cdots & \varepsilon_{fnl,ng} \\ \varepsilon_{f2,1} & \varepsilon_{f2,2} & \cdot & \varepsilon_{fnl,ng} \\ \cdot & \cdot & \cdot & \cdot \\ \cdot & \cdot & \cdot & \cdot \\ \varepsilon_{fnl,1} & \cdot & \cdot & \varepsilon_{fnl,ng} \end{bmatrix} \begin{bmatrix} P_1 \\ P_2 \\ \cdot \\ P_{ng} \end{bmatrix} \tag{8.29}$$

Flow reserve for eth line is defined as,

$$P_{fre} \triangleq \frac{P_{fe}^{\max} - P_{fe}}{P_{fe}^{\max}} \tag{8.30}$$

That is,

$$\begin{bmatrix} P_{fr1} \\ P_{fr2} \\ . \\ . \\ P_{frnl} \end{bmatrix} = \begin{bmatrix} 1 \\ 1 \\ . \\ . \\ 1 \end{bmatrix} - \begin{bmatrix} \frac{1}{P_{f1}^{\max}} & 0 & 0 & 0 \\ 0 & \frac{1}{P_{f2}^{\max}} & 0 & 0 \\ 0 & 0 & 0 & 0 \\ 0 & 0 & 0 & \frac{1}{P_{fnl}^{\max}} \end{bmatrix} \begin{bmatrix} \varepsilon_{f1,1} & \varepsilon_{f1,2} & \cdots & \varepsilon_{fnl,ng} \\ \varepsilon_{f2,1} & \varepsilon_{f2,2} & . & \varepsilon_{fnl,ng} \\ . & . & . & . \\ . & . & . & . \\ \varepsilon_{fnl,1} & . & . & \varepsilon_{fnl,ng} \end{bmatrix} \begin{bmatrix} P_1 \\ P_2 \\ . \\ P_{ng} \end{bmatrix} \tag{8.31}$$

Or,

$$\begin{bmatrix} P_{fr1} \\ P_{fr2} \\ . \\ . \\ P_{frnl} \end{bmatrix} = \begin{bmatrix} 1 \\ 1 \\ . \\ . \\ 1 \end{bmatrix} - \begin{bmatrix} \frac{1}{P_{f1}^{\max}} & 0 & 0 & 0 \\ 0 & \frac{1}{P_{f2}^{\max}} & 0 & 0 \\ 0 & 0 & 0 & 0 \\ 0 & 0 & 0 & \frac{1}{P_{fnl}^{\max}} \end{bmatrix} \begin{bmatrix} \varepsilon_{f1,1}P_1 & \varepsilon_{f1,2}P_2 & \cdots & \varepsilon_{f1,ng}P_{ng} \\ \varepsilon_{f2,1}P_1 & \varepsilon_{f2,2}P_2 & . & \varepsilon_{f2,ng}P_{ng} \\ . & . & . & . \\ . & . & . & . \\ \varepsilon_{fnl,1}P_1 & . & . & \varepsilon_{fnl,ng}P_{ng} \end{bmatrix} \tag{8.32}$$

Total flow margin available in the system is,

$$\sum_{e=1}^{nl} P_{fre} = nl - \left\{ \frac{P_1}{P_{f1}^{\max}} \sum_{e=1}^{ng} \varepsilon_{fe,1} + \frac{P_2}{P_{f2}^{\max}} \sum_{e=1}^{ng} \varepsilon_{fe,2} + \cdots + \frac{P_{ng}}{P_{fnl}^{\max}} \sum_{e=1}^{ng} \varepsilon_{fe,ng} \right\} \tag{8.33}$$

Maximization of (8.33) can be stated as,

$$\begin{gathered} \max J = \max \sum_{e=1}^{nl} P_{fre} = nl - \{c_1 P_1 + c_2 P_2 + \cdots + c_{ng} P_{ng}\} \\ c_1 = \frac{1}{P_{f1}^{\max}} \sum_{e=1}^{ng} \varepsilon_{fe,1};\, c_2 = \frac{1}{P_{f2}^{\max}} \sum_{e=1}^{ng} \varepsilon_{fe,2};\, \cdots;\, c_{ng} = \frac{1}{P_{fnl}^{\max}} \sum_{e=1}^{ng} \varepsilon_{fe,ng} \end{gathered} \tag{8.34}$$

subject to,

$$P_1 + P_2 + \cdots + P_{ng} = K(a\ constant) \tag{8.35}$$

With coefficients defined in (8.34), the LP can be solved in a manner similar to that described above.

8.4.1 Example

For the system in Fig. 8.1, let,

$$P_{f1}^{\max} = 100\,\text{MW},\ P_{f2}^{\max} = 100\,\text{MW},\ P_{f3}^{\max} = 50\,\text{MW},\ P_{f4}^{\max} = 30\,\text{MW},$$
$$P_{f5}^{\max} = 60\,\text{MW},\ P_{f6}^{\max} = 50\,\text{MW},\ P_{f7}^{\max} = 50\,\text{MW}.$$

Capacities of the generators are given as $P_1^{\max} = 100\,\text{MW}$ and $P_2^{\max} = 75\,\text{MW}$.
Power flow fractions for lines are obtained from (4.20) and arranged in matrix form as,

$$\begin{bmatrix} \varepsilon_{f1,1} & \varepsilon_{f1,2} \\ \varepsilon_{f2,1} & \varepsilon_{f2,2} \\ \varepsilon_{f3,1} & \varepsilon_{f3,2} \\ \varepsilon_{f4,1} & \varepsilon_{f4,2} \\ \varepsilon_{f5,1} & \varepsilon_{f5,2} \\ \varepsilon_{f6,1} & \varepsilon_{f6,2} \\ \varepsilon_{f7,1} & \varepsilon_{f7,2} \end{bmatrix} = \begin{bmatrix} 0.7221 & -0.1202 \\ 0.2778 & 0.1202 \\ 0.1268 & 0.2006 \\ 0.1488 & 0.2089 \\ 0.3085 & 0.3448 \\ 0.1295 & 0.0481 \\ 0.0422 & 0.0156 \end{bmatrix}$$

Power-flow reserves are given by (8.31), i.e.,

$$\begin{bmatrix} P_{fr1} \\ P_{fr2} \\ \cdot \\ \cdot \\ P_{frnl} \end{bmatrix} = \begin{bmatrix} 1 \\ 1 \\ \cdot \\ \cdot \\ 1 \end{bmatrix} - \begin{bmatrix} 1 & & & & & & \\ & 1 & & & & & \\ & & \frac{1}{0.5} & & & & \\ & & & \frac{1}{0.3} & & & \\ & & & & \frac{1}{0.6} & & \\ & & & & & \frac{1}{0.5} & \\ & & & & & & \frac{1}{0.5} \end{bmatrix} \begin{bmatrix} 0.7221 & -0.1202 \\ 0.2778 & 0.1202 \\ 0.1268 & 0.2006 \\ 0.1488 & 0.2089 \\ 0.3085 & 0.3448 \\ 0.1295 & 0.0481 \\ 0.0422 & 0.0156 \end{bmatrix} \begin{bmatrix} P_1 \\ P_2 \end{bmatrix}$$

Objective function (8.34) is,

$$J = 7 - \{2.6071P_1 + 1.7997P_2\}$$

To maximize this function we must *first* use maximum capacity of the generator which best influences the maximization of the objective function, i.e., generator P_2. Next, we utilize full capacity of the generator with next highest coefficient, and so on. Solution therefore is,

$$P_2 = 75\,\text{MW and}\, P_1 = 95\,\text{MW}.$$

Maximum power-flow reserve capacity is then,

$$\begin{aligned} J &= 7 - \{2.6071 * 0.95 + 1.7997 * 0.75\} \\ &= 3.1731\,\text{pu} \end{aligned}$$

With injections of 130 and 40 MW, the reserve capacity is,

$$\begin{aligned} J &= 7 - \{2.6071 * 130 + 1.7997 * 40\} \\ &= 2.8904\,\text{pu} \end{aligned}$$

Effect of rescheduling generations on reserve margins in transmission lines is clearly seen from the above result. This optimization can be employed to *relax* line flows in the network.

8.5 Tieline Dispatch

In deregulated systems, generators inject power into the system as per contracts and schedules. Profit being the main consideration, a generation company per se is not interested in economic operation of the overall system which includes *other* owners. Relevance of global economic load dispatch is therefore lost. Let a system export $P_t + jQ_t$ over a tieline. We substitute tieline power (supposed to be maintained constant) as load which is included as impedance in the network. We shall assume given condition to be an optimal-use operating point. When an adjoining area requires more power over tieline, the generators in the supplying area need to adjust to another point of optimal use. Allocating generator powers to achieve this is called is *tieline dispatch*. We explain our method with an example.

8.5.1 Example

Let Fig. 8.1 represent the system having two different owners of generator 1 and 2. Let the 40 MW shown at bus 4 represent the *tieline power—export—*from this system to other adjoining system (not shown on the figure). The export is to be maintained at constant level of 40 MW at all times. *Find the generator injection/s that will achieve this.*

We continue our solution from the result of optimal use obtained in Sect. 8.2.1. For optimal use the generator 2 supplies 75 MW (maximum capacity) and generator 1, 95 MW (maximum capacity is 100 MW). With these injections the load consumptions are as shown below.

$$\begin{bmatrix} p_{Ld1} \\ p_{Ld2} \\ p_{Ld3} \\ p_{Ld4} \end{bmatrix} = \begin{bmatrix} 0.1213 & 0.1249 \\ 0.2636 & 0.2667 \\ 0.2335 & 0.2371 \\ 0.3440 & 0.3525 \end{bmatrix} \begin{bmatrix} P_1 \\ P_2 \end{bmatrix} = \begin{bmatrix} 0.1213 & 0.1249 \\ 0.2636 & 0.2667 \\ 0.2335 & 0.2371 \\ 0.3440 & 0.3525 \end{bmatrix} \begin{bmatrix} 95 \\ 75 \end{bmatrix} = \begin{bmatrix} 20.8910 \\ 45.0445 \\ 39.965 \\ 59.1175 \end{bmatrix}$$

We see that tieline power is 39.965 and, not 40 i.e., a value which is required to be maintained constant. In optimal use LP, generator 2 was allocated its maximum capacity since it had maximum of its fraction consumed in loads. The shortfall of 0.045 therefore will have to be met by the generator 1 (the last generator to be allocated power in LP, whose maximum has not yet been reached). Power consumption due to generator 1 in this load (tieline at bus 4) is analytically equal to $0.2335P_1$. Therefore an additional power must be supplied by generator 1.

$$P_1^{additional} = \frac{0.035}{0.2335} = 0.1499 \text{ MW}$$

Injection of generator 1 *now* is 95.1499. Since this generator has contribution in other loads too, the consumptions in other loads are also modified, i.e.,

$$\begin{bmatrix} p_{Ld1} \\ p_{Ld2} \\ p_{Ld3} \\ p_{Ld4} \end{bmatrix} = \begin{bmatrix} 0.1213 & 0.1249 \\ 0.2636 & 0.2667 \\ 0.2335 & 0.2371 \\ 0.3440 & 0.3525 \end{bmatrix} \begin{bmatrix} 95.1499 \\ 75 \end{bmatrix} = \begin{bmatrix} 20.9092 \\ 45.0840 \\ 40.00 \\ 59.1619 \end{bmatrix}$$

Normally, shortfall in tieline power should be met by the generator whose *incremental fuel cost* is lowest. With more generators and capacity of the *lowest* having limits, the above problem assumes an LP format and can be solved using linear programing.

8.6 State Estimation

Measured values of variables contain errors and deviate from their true values. These measurements are used in an optimization procedure to find an *estimate* of the state variables (voltages and phase angles) of the system. The objective is to minimize norm of residual errors i.e., difference between measurements and the true values of variables. Usual bus loading equations are used. An implicit assumption in state estimation is that more the number of measurements, the better is the accuracy. A large number of measurements are therefore used to obtain a *good* state estimate. State estimation procedures are however prone to problems of unobservability, ill-conditioning and non-convergence. Using a large number of measurements in the hope of getting a better estimate is fraught with unknown consequences due to grossly erroneous measurements sneaking into calculations. Data are therefore required to be cleaned with the help of gross error detection algorithms. Modular approach starts with measurements at generating stations whose correctness is ensured at the outset, by duplicating them, if required.

These accurately measured injections are used in MLF to obtain an exact solution of the network in closed form. The question of *estimation* simply does not arise. Only accuracy of the measurements at generating stations needs to be ensured. Investment required for duplicating measurements is considerably more economical than the overall deployment of PMUs which require extensive data analytics with sophisticated algorithms making them susceptible to errors. Accuracy of MLF is linked only to the accuracy of the measured injections.

8.7 Conclusion

Network solutions and optimization of power systems involve solution of partial differential equations. Various iterative techniques are required to be used for this purpose. Modular approach offers, for the first time, a formulation of the problem that employs a set of *linear* equations. This is a simpler exercise than those using ILF. Assumption of normal voltages at buses is generally admissible. Use of constant impedances for loads is therefore justified. Optimization procedure using the concept of MLF is very simple. The advent of deregulation has made it desirable for each owner of generation to have an estimate of how much revenue his investment can generate. MLF based procedures can be used to advantage in usage based pricing. Modular approach offers possibilities of further investigation on economics of restructured power systems.

Reflections

Where does the 'economic dispatch' stand in the market-dominated power systems? With dispatches decided by the markets, we do not seem to be as much concerned about fuel costs as in the early days of non-deregulated systems. With restructured, market-controlled power systems, coordination equations seem useless and the role for incremental fuel costs in grid operation not significant anymore! Fuel cost is only a component of the net marginal cost. The thrill of economic dispatch formulation goes for a toss. It seems to be useful only for Gencos owning multiple generators. Concept of optimal use introduced in this chapter is concerned with best use of injected powers, a relevant concept in deregulated systems.

References

1. Wood, A.J., Wollenberg, B.F., Sheble, G.B.: Power Generation and Control. John Wiley and Sons (1996)
2. Stagg, G.W., El-Abiad, A.H.: Computer Methods in Power System Analysis, 1st edn. McGraw-Hill Inc., US (1968)

Chapter 9
Blackout Incipient

The educational value lies in the experience of following a line of reasoning to its logical conclusion; and in the thrill of suddenly realizing that our topic has thereby gained a much wider significance than initially seemed possible.

Ernst A. Guillemin

9.1 Introduction

System studies may be classified into three categories; normal, diagnostic and post-mortem. Normal studies include load flow, economic dispatch, optimization, etc. Diagnostic studies, usually called contingency analyses are conducted to predict effect of line outages. Post-blackout investigation is post-mortem which includes studies probing whys of a system failure. Blackouts can be triggered by line or generator outages in vulnerable regime. Objectives of post-mortem analyses are to identify the causes of failures and make recommendations so that outages do not occur in future. This learning approach is expensive in terms of cost and time and is generally case-specific and system-specific. Blackouts also do occur due to human limitations. Critical control actions are occasionally not taken by operators due to lack of understanding of its criticality or other considerations. During over-stressed conditions of the system, this inaction may trigger equipment or line outages with cascading failures resulting in blackouts. Deregulated or decentralized systems comprising interconnected regions are operated with commercial objectives and regional interests, overall security being considered a lower priority. Such systems are therefore more vulnerable to blackouts as compared to vertically integrated systems.

M.V. Hariharan et al., *Modular Load Flow for Restructured Power Systems*,
Lecture Notes in Electrical Engineering 374, DOI 10.1007/978-981-10-0497-1_9

9.2 Limitations in Blackout Related Studies

In contingency analyses distribution factors are used to find congested lines after outages [1]. A global load flow is a pre-requisite for calculation of distribution factors. Iterative load flow is ill-suited for this purpose as specifying PV buses and their voltages is difficult (since many PV buses would have switched to PQ type). If this decision is not taken judiciously, non-convergence will most likely occur. Difficulties in conducting global load flow may therefore stall the analysis. Distribution factors are also known to vary with choice of the slack bus which is unclear to the analyst. With severally owned generators in geographically scattered areas, as in large grids, this problem is not easy to deal with. In fact the results can be grossly erroneous. MLF does not need specification of slack or PV buses and therefore has definite advantage.

9.3 Blackout-Incipient and Vulnerability

When small changes in the network result in disproportionately large changes in system conditions the system is said to be vulnerable. *Blackout-incipient* characterizes a vulnerable state and is usually caused by gradually increasing load in some part of the grid. Purpose of blackout incipient study (BIS) is to find answers to questions: In what manner do the line flows, voltages and angular separations evolve in the incipient? At what load levels would the limits for these variables be reached and would cause relay operations (outages)? MLF can provide unambiguous answers to these questions. Iterative load flow study is ill-suited for this purpose. Generating station data from SCADA can be readily used in MLF. *Contingency analysis* examines the system for 'outages' only and is therefore different from BIS. Blackout incipient study monitors system's health continuously as if it is under intensive care and assumes paramount importance. Early warning systems can be designed using BIS.

To conduct BIS, the system has to be first suitably benchmarked. Detailed network (if available) or an equivalent can be used. Benchmarking is done by comparing simulation results with available measurements. Judicious parameter adjustments need to be made to bring about a match between the two. We illustrate BIS on an equivalent of the NEW (North-East-West) grid using antecedent conditions of massive blackouts that occurred in the Indian grid on 30th July 2012 and 31st July 2012. A brief description of the Indian Grid and the blackout events is given in the following sections.

9.4 The Indian Grid—Overview

Electric grids are formed by interconnection of network clusters through transmission lines. Each cluster has its own generation and load. Tie lines enable flow of power from surplus regions to deficit regions. Such flows help increase system reliability. Interconnection of smaller grids results in larger grids, which in turn may form still larger grids. In the Indian context each state has an electrical grid confined to its geographical area. State grids are interconnected to form regional grids. Regional grids in turn form National Grid. The term "NEW Grid" is collectively used to describe the group of Northern, Eastern and Western regional grids. In the NEW grid, WR, ER and NER are surplus regions. The structure of the Indian National Grid along with its regional constituents is shown in Fig. 9.1

Before December 2013 India had two asynchronously connected main grids: Northern Grid and the Southern Grid connected to the former by two HVDC lines. Some detailed and interesting information about the Regional Grids are given in Table 9.1.

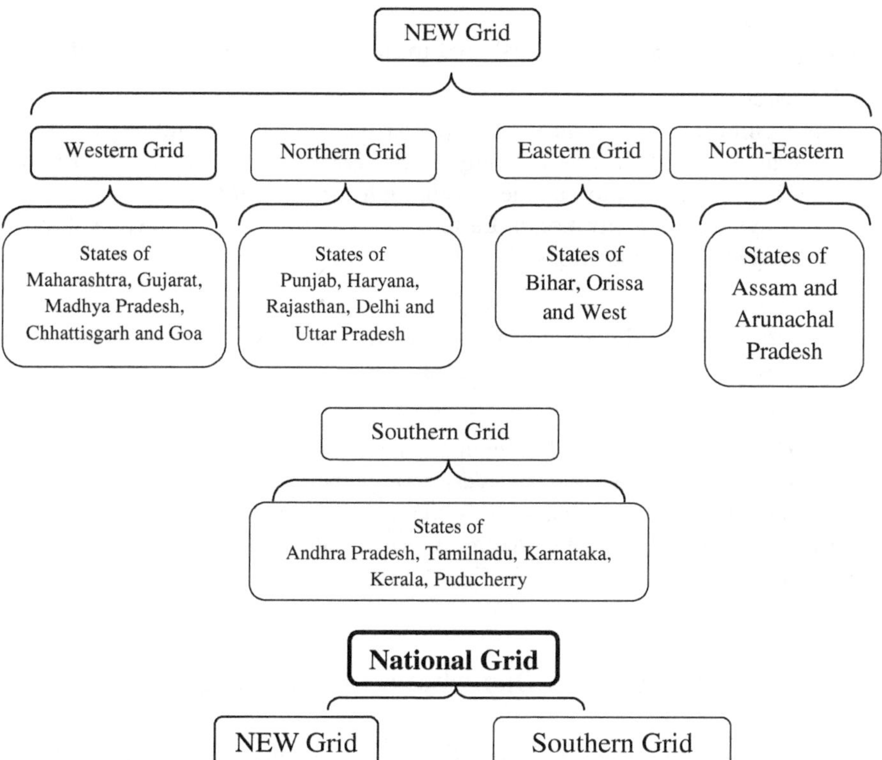

Fig. 9.1 Structure of the Indian grid

Table 9.1 Salient details of regional grids [2]

Item		Northern grid	Eastern grid	N. E. grid	Western grid	Southern grid
Installed capacity (MW)	July 2012	56,089	26,838	2455	67,636	53,362
	July 2015	73,202	33,321	3513	100,137	65,686
Increase in 3 years		31 %	24 %	43 %	48 %	23 %
Details as on July 2015	Thermal	46,275	28,773	2009	77,844	36,723
	Hydro	17,798	4114	1242	7447	11,398
	Nuclear	1620	NIL	NIL	1840	2320
	Renewables	7512	436	762	13,006	15,245

Some general remarks about the various grids are given below [3].

Northern Grid
Major generating stations including super thermal power stations (STPS) at Rihand and Singrauli are located in the eastern part of the NR grid and major load centers are in central and western part of the grid. This results in bulk power transmission from eastern to western part over long distances.

Eastern Grid
Surplus power from Bhutan is transferred to India through the Eastern grid.

North-Eastern Grid
There is significant increase in thermal power generation since July 2012. North Eastern grid is directly connected only to the Eastern grid. The power transfer from/to other regions has to be wheeled through the Eastern Grid.

There is no nuclear generation in Eastern and North-Eastern Grid while all other regions have nuclear generations.

Western Grid
The increase in installed capacity in this region since July 2012 is basically from renewable sources.

Southern Grid
Two HVDC links connect the Southern Regional Grid and the rest of the grids in the North, East, North-East and the Western regions. In December 2013, AC lines were installed to provide a synchronous connection between the southern grid and the rest of the country. Thus the dream of an All-India Grid had become a reality.

The NEW Grid map is shown in Fig. 9.2.

9.5 Case Study 1: Grid Disturbance on 30th July 2012

A major grid disturbance at 0233 h on 30th July caused blackout in almost all parts of northern region. There was large demand from Northern grid and it was drawing from both Western and Eastern grids. The Western grid was under-drawing and

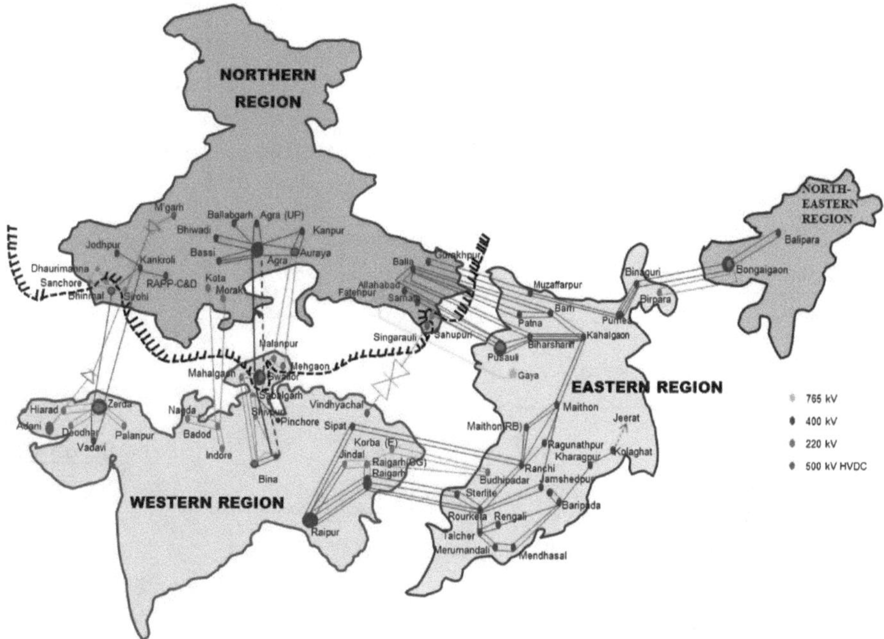

Fig. 9.2 NEW grid with tie lines [4]

some major transmission lines were out either due to planned outage or forced outage. As demand–supply balance among the regions was highly skewed; it led to a stressed system. Inter-regional flows at 0200 h, half an hour before the onset of the blackout are shown in Fig. 9.3 (reproduced from [4, p. 7]). With operating conditions shown in the figure the system is on the brink of a collapse as characterized by over-loading of transmission lines and low voltages resulting from reactive powers hitting the capability limits of generators.

9.6 Load Representation

In BIS, loads are represented by constant impedances evaluated from available data. Constant impedance indicates that the load power varies nonlinearly with square of the applied voltage. Constant PQ assumption as in iterative load flow procedures loses validity as the system gets more and more over-stressed. We also assume that generators supply reactive power, say, 30 % of real power. Similarly each load MVA is assumed to consist of reactive power component of 30 % of its active power (power factor of 0.95). These values are usually available from SCADA and would really form data, not assumptions.

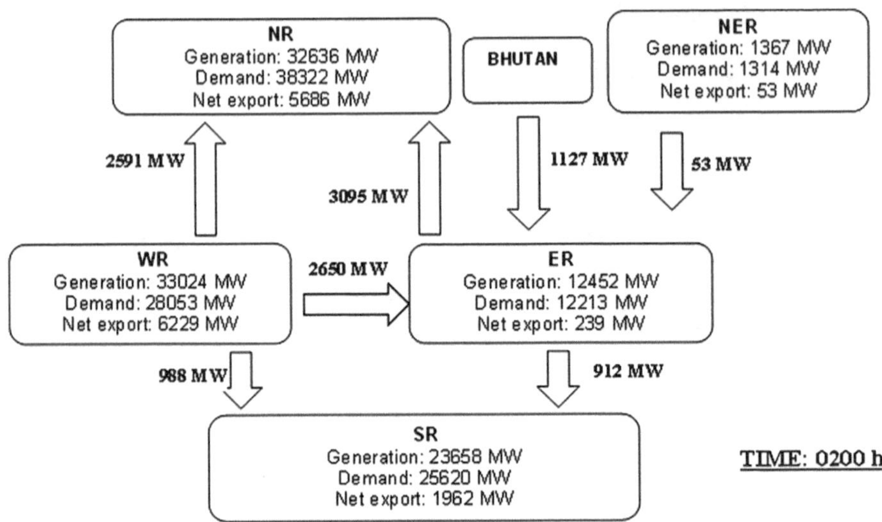

Fig. 9.3 Antecedent conditions on 30th July 2012 (0200 h)

9.7 Radial-Mesh Representation for Sub-grids

Radial-Mesh (RM) representation consists of radial lines and a ring network which tries to capture nodal as well as distributive features of each region (Fig. 9.4). The RM equivalent is obtained from generations and tie line data. Total load of the region is equally distributed among the boundary buses of the region. All generation is assumed to be located at a central bus. For example, WR has total load of 28,053 MW. It has 3 boundary buses numbered 6, 11 and 12. Load assigned to each is thus 9351 MW. In Fig. 9.3, for WR, generation does not sum up to demand and net export. Hence generation of 34,282 MW is taken which is sum of WR's total load and net export. Generation of 34,282 MW is at the radial center bus—the generator bus numbered 1. Transmission line elements exist between each boundary node and the generator bus, and also between a boundary bus and the two adjacent buses (Fig. 9.4). The inter-regional radial lines and those between boundary buses are assigned typical line impedances of $0.001 + j0.004$ pu on 1000 MVA base. Charging admittances of 0.0001 are added to all lines. Tie line impedances are taken to be smaller by an order and their charging admittances are assumed a little higher, 0.005 pu. Some adjustment of tie line impedances was required to match the measured line flows with the load flow values. From the given antecedent conditions, each region of interconnected grid is represented as a radial-mesh system. Southern region is asynchronously connected to the Northern grid and is represented as load equal to total power flowing from Western region and Eastern region to Southern region. Modular Load Flow results on this model of NEW grid match the measured power flows on tie lines at 0200 h on 30th July. Single line diagram for the NEW grid is shown in Fig. 9.5.

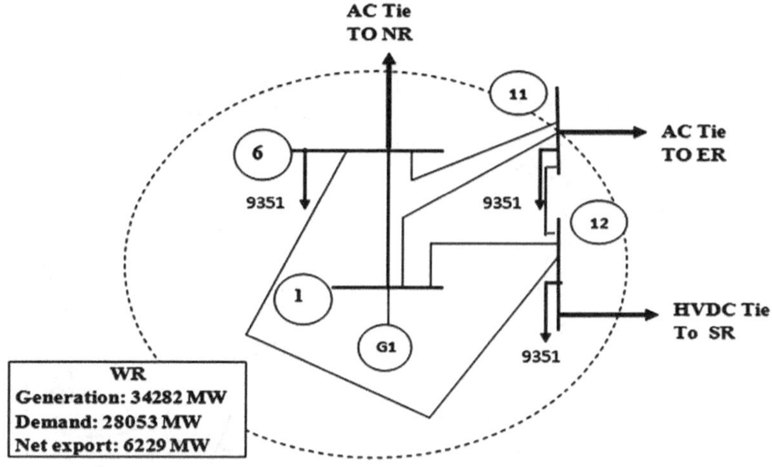

Fig. 9.4 Radial-mesh network for region 1 (WR)

9.8 Benchmarking

For benchmarking, MLF is conducted with RM equivalents for generations and tie line flows on 30 July at 0200 h. This is the starting point of BIS. MLF results are matched with the pre-outage tie line flows by adjusting a few line parameters. With benchmarking done, r*elative* variation in line flows and bus voltages for incremental changes in loads are obtained using MLF for each incremental change. The flows and voltages when checked against respective relay settings determine the *sequence* of tripping. Load in the Northern region is varied for the BIS. Table 9.2 shows transmission capacity of inter-regional corridors as on 31.3.2012.

9.9 Algorithm

Step 1: *Obtain values of injected real and reactive powers from measurements at generating stations.*

Step 2: *Calculate tie line flows and boundary bus voltages using MLF. i.e.,* (4.19) − (4.22) *and,* (4.25) *respectively, starting with load of* 19,161 MW *at load buses in NR* (0200 h, 30 July 2012).

Step 3: *Increment load in the NR (approximately* 4 *%—see* Table 9.3).

Step 4: *Compute new Z-bus.*

Step 5: *Find new flows and voltages using equations mentioned in step* 2.

Step 5: *Repeat steps* 2*–*4 *for a sufficient range.*

Figure 9.6, reproduced from [4], shows tie line flows at 0230 h on 30th July. Table 9.4 shows measured and calculated flows from MLF on tie lines at 0200 h and

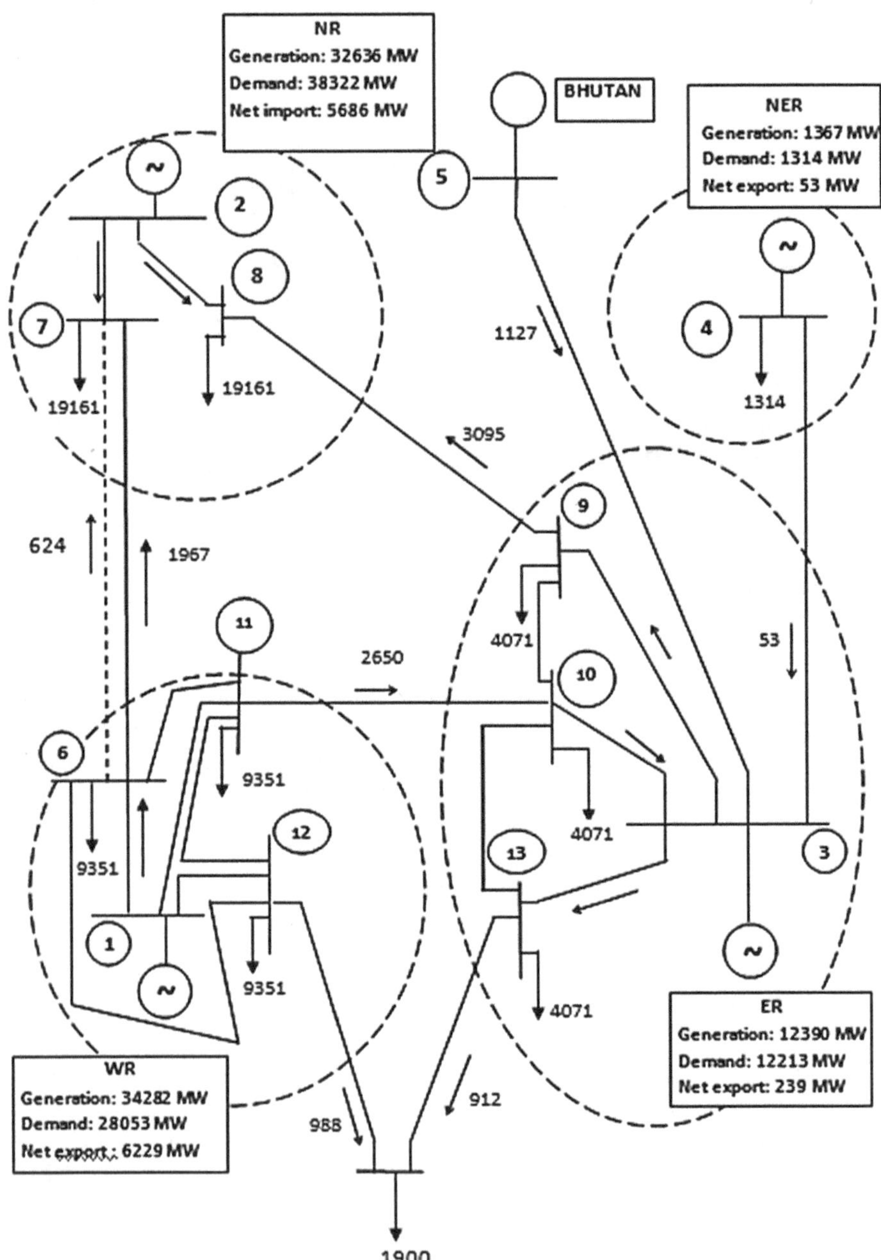

Fig. 9.5 SLD for the NEW grid with conditions at 0200 h

Table 9.2 TTC and transmission capacity of inter-regional corridors

Corridor	Transmission capacity (as on 31.3.2012)	TTC (highest during 2011–12)	% of TTC to transmission capacity	Capital investment made in XI plan (in crore)	% of total investment
WR-NR	4220	2000	47	465	11
WR-ER	4390	1000	23	1009	24
ER-NER	1260	500	40	–	–
WR-SR	1520	1000	66	–	–
ER-NR	10,030	4200	42	2706	63
ER-SR	3630	2830	78	106	2
Total	25,050	11,530	–	4286	100

Source [5]

Table 9.3 Loading in the NR region (approximate steps of 4 %)

Load at load buses in NR (MW)	WR-NR tie line (MW)	ER-NR tie line (MW)	WR-ER tie line (MW)	ER-NER tie line (MW)
19,161 (Load at 0200 h)	2591.9	3095	2650.5	53.4
19,935	2964.5	3407.1	2729.6	55.5
20,741	3337.1	3718.2	2807.8	56.6
21,579	3708.9	4027.4	2884.9	58.2
22,451	4079.8	4334.8	2960.7	59.8
22,900	4264.8	4487.6	2998.1	60.6

Table 9.4 Power flows on inter-regional tie lines at 0200 h and 0230 h on 30th July

Tie line	From bus	To bus	Power flows (MLF), MW at 0200 h	Power flows (measurement), MW at 0200 h	Power flows (MLF), MW at 0230 h	Power flows (measurement), MW at 0230 h
WR-NR	6	7	2591.9	2591.0	2629.6	2636.0
NR-ER	8	9	−3095.0	−3095.0	−3126.7	−3120.0
WR-ER	11	10	2650.5	2650.0	2658.5	2650.0
NER-ER	4	3	53.4	53.4	53.6	95.0
Bhutan-ER	5	3	1125.7	1127.0	1125.7	

0230 h. Figures 9.3 and 9.6 and Table 9.4 show calculated and measured tie line flows match, which validate our model. Benchmarking of the model is therefore confirmed. For approximately 20 % load change at load buses in NR, WR-NR tie line hits its maximum capacity. Plots of tie line flows as percentage of their limits and, boundary-bus voltages are shown in Table 9.5 and Figs. 9.7 and 9.8. It is seen that tie line flows increase and the voltages decline all over the system. It is during this period that limits of relay setting are likely to get violated. Depending on their settings,

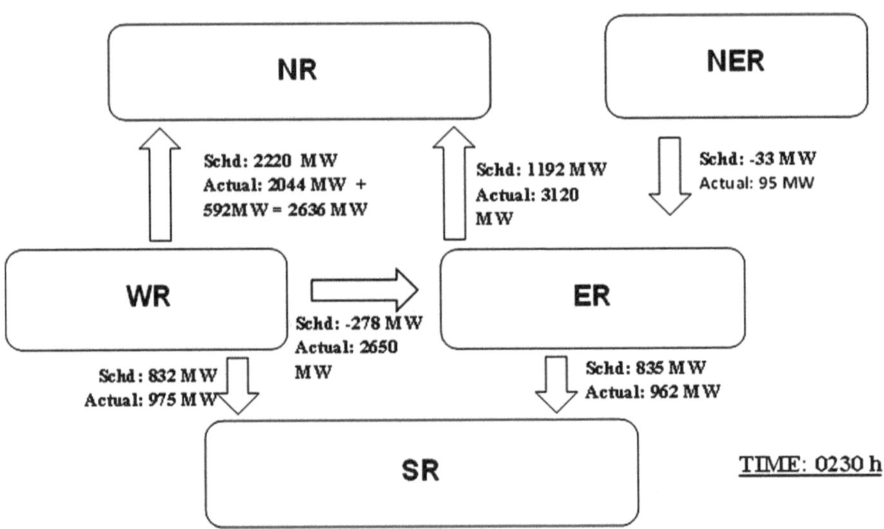

Fig. 9.6 Power flow on Inter-regional links

Table 9.5 Percentage power flow with respect to maximum capacity of tie lines

Change in load in NR	WR-NR tie line (%)	ER-NR tie line (%)	WR-ER Tie Line (%)	ER-NER Tie Line (%)
Load at 0200 h	61.41943	30.85743	60.37585	4.238095
4 % increase	70.24882	37.07079	63.959	4.404762
8 % increase	79.0782	40.15354	63.959	4.619048
12 % increase	87.88863	43.21834	65.71526	4.746032
16 % increase	96.67773	43.21834	67.44191	4.746032
20 % increase	101.0616	44.74177	68.29385	4.809524

sequence of breakdown of equipment and transmission lines and hence collapse of the system can be predicted. Control actions such as, load shedding, need to be taken urgently during this period to prevent outages of major component, such as a critical generator or a tie line. If and when that happens the system enters the cascading failure mode. Once that sets in, control action becomes humanly impossible.

Two hours before blackout, on 29th July, operators sent messages [4, p. 36]. But, for one reason or the other control action did not take place. For example, NR did not shed load in spite of the urgency to do so to ensure integrity of the system. Their argument before CERC lay in justification that no revision in schedule was intimated to them in time (planned outage of one of the tie lines between WR and NR, for maintenance, was not considered by operator to do this). WR did not control its under-drawal. With transmission capacity of corridor between WR and NR severely depleted, congestion occurred which led to snapping of WR-NR tie. Power to NR got routed through ER thereby overloading its transmission system and reducing

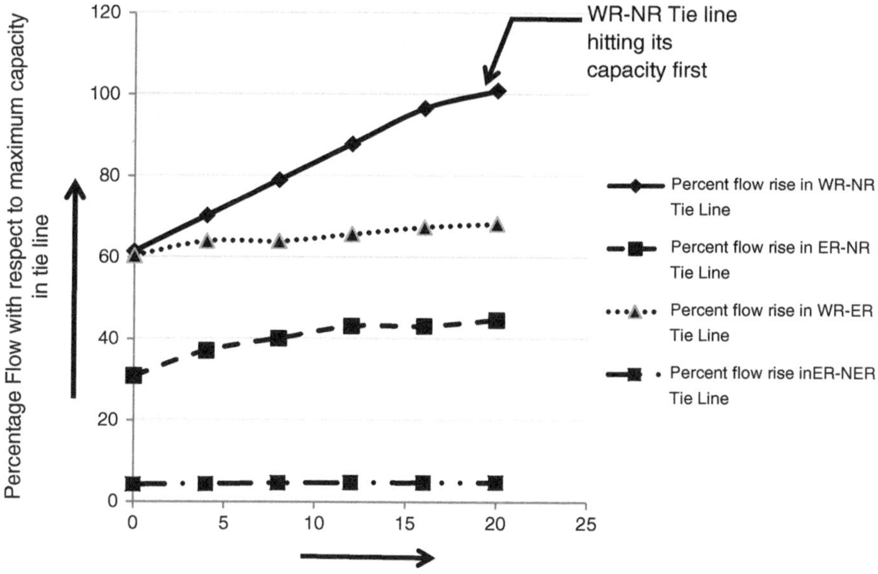

Fig. 9.7 Gradual increase in load in NR region

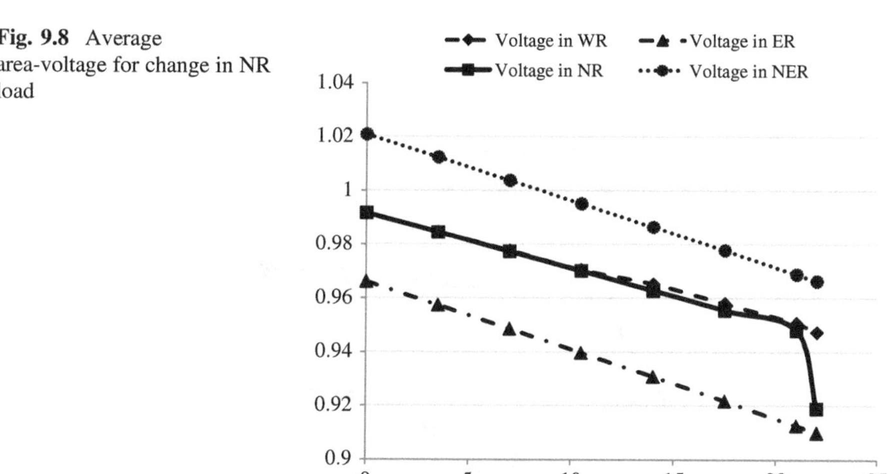

Fig. 9.8 Average area-voltage for change in NR load

voltages all over, except the WR. Trends plotted in Figs. 9.7 and 9.8 (X-axis shows % increase in load) might have helped the operator take better decisions.

It is observed that at 0200 h on 30th July, WR-NR tie line was loaded to 61.5 % of its total capacity. WR-ER tie line was loaded to 60.38 %. For increasing demand in the NR step increase of approximate 4 % was used. WR-NR corridor was the first to hit the maximum capacity and WR-NR tie line snapped. Corresponding tie line flows before and after tripping of WR-NR tie line are shown in Table 9.6.

Table 9.6 Tie line flows before and after tripping of WR-NR tie line

Tie line	From bus	To bus	MW flow, before tripping	MW flow, after tripping
WR-NR	6	7	4264.8	0 (tripped)
NR-ER	8	9	−4487.6	−4554.3
WR-ER	11	10	2998.1	3723.4
NER-ER	4	3	60.6	73.0
Bhutan-ER	5	3	1125.8	1126.0

Table 9.7 Angular separation in tie lines with increase in load in NR, rad

Percentage load change in NR	WR-NR	ER-NR	WR-ER
0	0.9346	1.6042	1.0035
4	1.2103	2.4665	0.9956
8	1.4409	2.5157	0.9871
12	1.6451	2.5659	0.978
16	1.8321	2.6171	0.9682
20	1.9208	2.6431	0.963

Table 9.7 and Fig. 9.9 show there is rapid angular separation in WR-NR tie line. Angular separation in ER-NR also is increasing but change is gradual as compare to steep rise in WR-NR tie line.

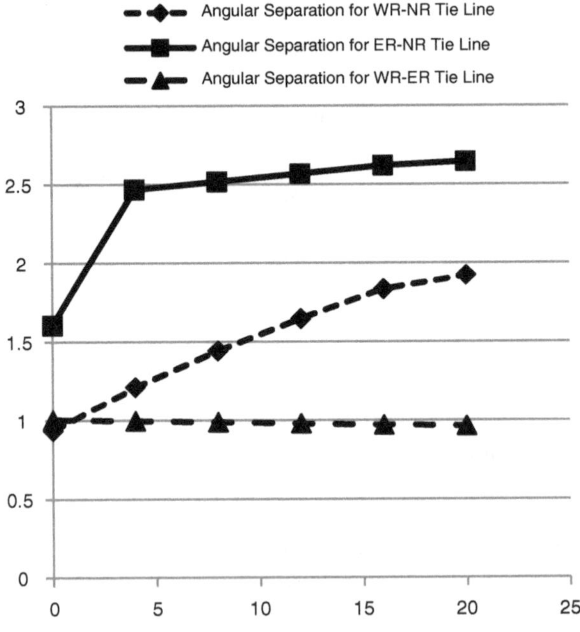

Fig. 9.9 Angular separation (rad) of tie lines with increase in load in NR on 30th July

9.10 Case Study 2: Grid Disturbance on 31st July 2012

The power was partly restored after a major disturbance on the night of 30th July that had caused most of the northern region plunge into darkness. Another disturbance occurred at 1300 h on 31st July which affected all regions of NEW grid except Western region. The Southern region survived on both occasions as it was asynchronously connected to rest of the grid. Antecedent conditions are schematically displayed in Figs. 9.10 (1230 h) and 9.11 (1257 h) [4]. Single line diagram for the NEW grid with antecedent conditions (at 1230 h) is shown in Fig. 9.12.

It is seen from Table 9.8 that power flows on tie lines are very close to power flows of snapshots at 1200 and 1257 h. Benchmarking is thus again vindicated. Effect of NR loading on tie line flows is shown in Table 9.9. Percent loadings (with respect to limits) are more meaningful from the point of view of relay operations and are shown in Table 9.10 and Fig. 9.13.

On 31st July, half an hour before cascade tripping, tie lines loading were 54 % for WR-NR and 32.95 % for WR-ER corridor. When WR-NR line trips, immediately most of power is routed through WR-ER tie line (Table 9.11). This causes WR-ER tie line also hitting its maximum capacity (4390 MW) resulting in tripping of this line. Effect on tie line of these sequential tripping is shown in Table 9.11. This was the reason most of the NEW grid except Western region witnessed blackout. Voltage profile is shown in Fig. 9.14.

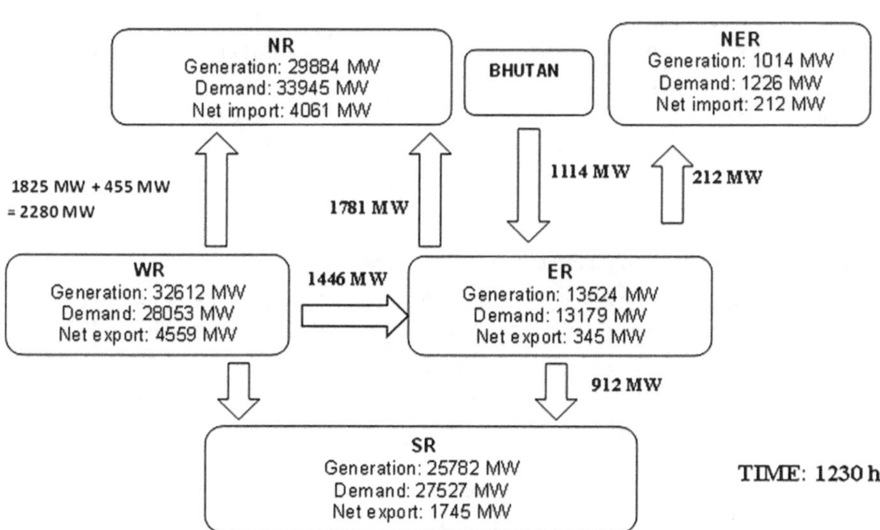

Fig. 9.10 Antecedent conditions on 31st July 2012 (1230 h)

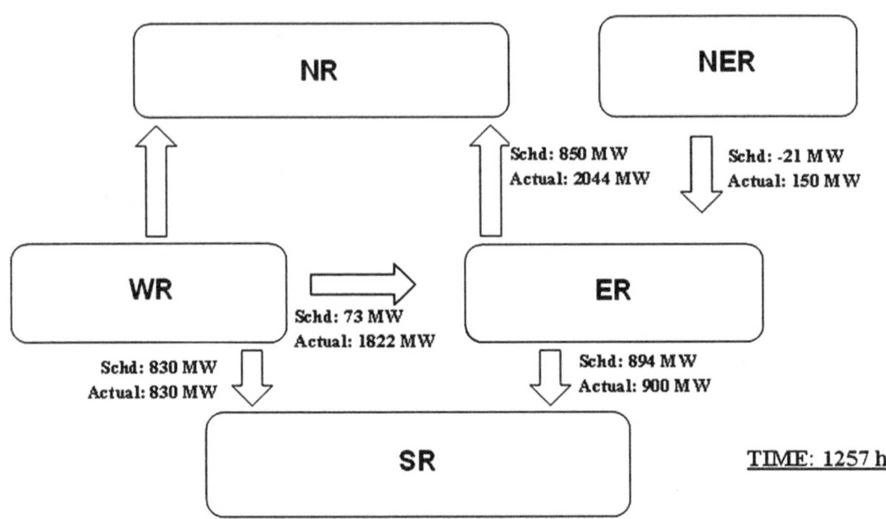

Fig. 9.11 Antecedent conditions on 31st July 2012 (1257 h)

Table 9.8 Power flows on inter regional tie lines at 1230 and 1257 h on 31st July 2012

Tie line	From bus	To bus	Power flows (MLF), MW at 1200 h	Power flows (measurement), MW at 1200 h	Power flows (MLF), MW at 1257 h	Power flows (measurement), MW at 1257 h
WR-NR	6	7	2280.1	2280.0	2588.4	2410.0
NR-ER	8	9	−1781.2	−1781.0	−2045.1	−2044.0
WR-ER	11	10	1446.6	1446.0	1489.0	1822.0
NER-ER	4	3	−212.0	212.0	−191.7	−150.0
Bhutan-ER	5	3	1112.7	1114.0	1112.8	–

Table 9.9 Load in NR region increased in steps of approximate 3.5 %

Load at load buses in NR (MW)	WR-NR tie line (MW)	ER-NR tie line (MW)	WR-ER tie line (MW)	ER-NER tie line (MW)
16973 (load at 1230 h)	2280.1	1781.2	1446.6	212
17,567	2588.4	2045.1	1489	191.7
18,182	2897.4	2309.2	1531.5	171.4
18,818	3206.4	2573	1574	151.1
19,477	3516	2836.7	1616.7	130.8
20,159	3825.5	3099.9	1659.3	110.5
20,865	4134.8	3362.4	1701.9	90.2
21,074	4224.3	3438.2	1714.2	84.3

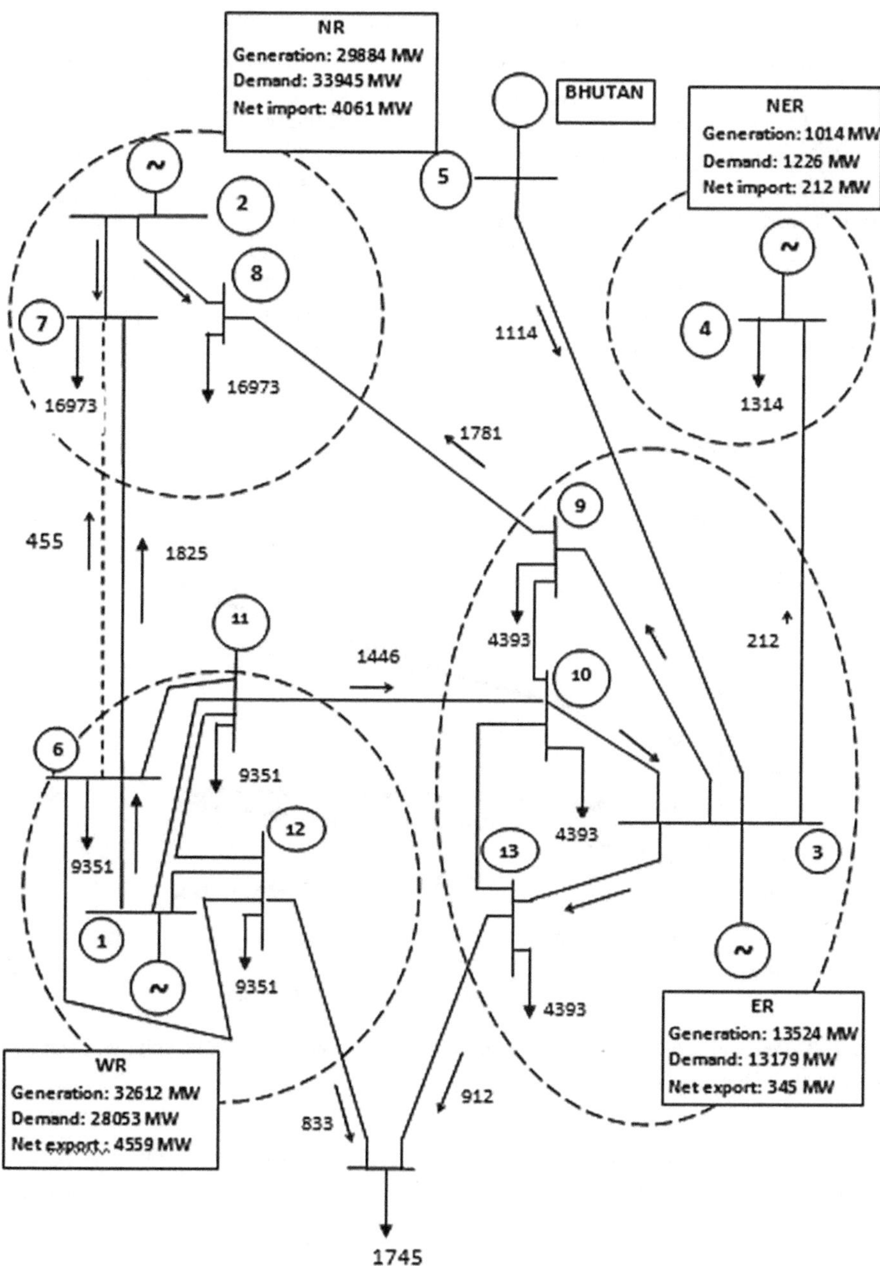

Fig. 9.12 Tie line flows on 31st July at 1230 h

Table 9.10 Percentage power flow with respect of maximum capacity of tie lines

Change in load in NR	WR-NR tie line (%)	ER-NR tie line (%)	WR-ER tie line (%)	ER-NER tie line (%)
Load at 1230 h	54.03081	17.75872	32.95216	16.8254
3.50 % increase	61.33649	23.02293	34.8861	13.60317
7 % increase	68.65877	25.65304	35.85421	11.99206
10.50 % increase	75.98104	28.28215	36.82688	10.38095
14 % increase	83.31754	28.28215	36.82688	10.38095
17.5 % increase	90.65166	30.90628	37.79727	8.769841
21 % increase	97.98104	33.52343	38.76765	7.15873
22 % increase	100.1019	34.27916	39.04784	6.690476

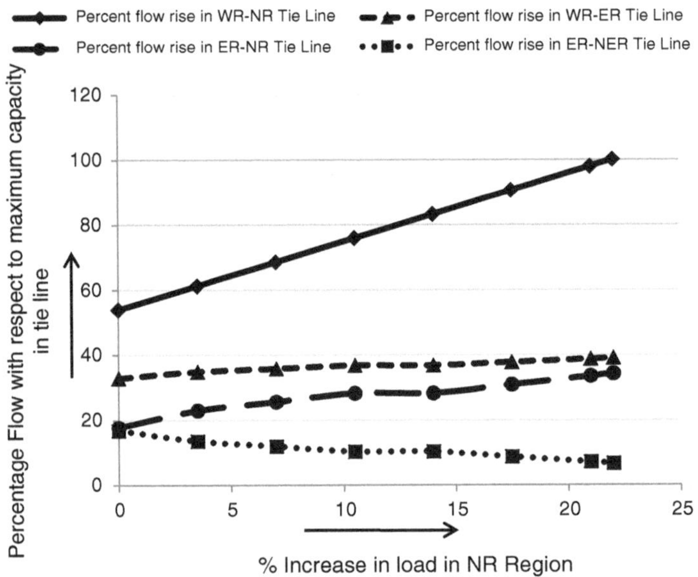

Fig. 9.13 Percentage flow with respect to maximum limits on tie lines, 31st July, % Increase in load in NR region

Table 9.11 Tie line flows after sequential tripping of WR-NR and then WR-ER

Tie line	From bus	To bus	Power flows MW, before WR-NR (6–7) trips	Power flows MW, after WR-NR (6–7) trips	Power flows MW, after 9–10 (a line in ER) trips
WR-NR	6	7	4224.3	0 (tripped)	0 (tripped)
ER-NR	8	9	−3438.2	−636.2	−3438.9
WR-ER	11	10	1714.2	7672.4	0 (tripped)
NER-ER	4	3	−84.3	135.9	34.0
Bhutan-ER	5	3	1112.9	1113.1	1112.9

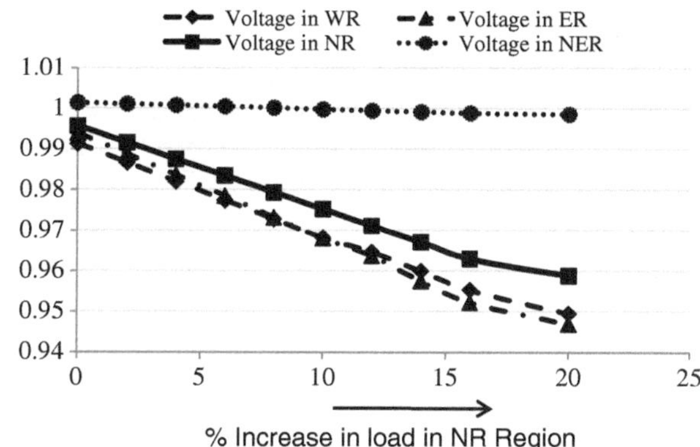

Fig. 9.14 Average area—voltages, 31st July, % increase in load in NR region

9.11 Conclusion

Blackout incipient study is not possible or extremely difficult with iterative load flow. It is shown in this chapter that using RM equivalents for sub-grids Modular Load Flow can be used to perform the blackout incipient study. Results may not exactly represent the actual conditions but are sufficiently indicative of trend of system's decline. Further research is needed to improve construction of RM networks. Forming an All-India grid has been dream of engineers and planners at the national and regional level in India. Similar efforts are being made in many countries. Plans for growth of generation, transmission and distribution emphasize the need for interconnection at local, state, regional and national levels keeping in mind the need for ensuring system stability, integrity and security. As systems expand, newer and faster methods of system analysis are needed to enable preventive actions to be taken. In this context, we believe that Modular Load Flow would play a very important role.

Reflections

Systems are sometimes forced to operate when generator voltages cannot be maintained constant. It is imperative to closely monitor the system during this critical period. Iterative load flows cannot be conducted for various reasons. MLF is like a shot in the arm to conduct such blackout incipient studies for developing early warning systems.

References

1. A.J. Wood, B.F. Wollenberg, Power Generation, Operation and Control (John Wiley and Sons, New York 1996)
2. Central Electricity Authority official website http://www.cea.nic.in/

3. Report of the Task Force on Technical Study in regard to Grid Stability, October 2014, New Delhi (available on the net)
4. Report on the Grid Disturbance on 30th July 2012 and Grid Disturbance on 31st July 2012
5. Report of CAG of India on Planning and Implementation of transmission projects by PGCIL and Grid Management by POSOCO Limited for the year ended, March 2013 Report # 18 of 2014. Ministry of Power Government of India

Chapter 10
Dirac Structures

Equations are just the boring part of mathematics. I attempt to see things in terms of geometry...

Stephen Hawking

There is more to Modular Load Flow than just manipulation of algebraic equations. Ideas from mathematical physics, function-vector spaces and graph theory are embedded in the Modular approach. Material in this chapter is meant for readers who want to delve into depth and understand nuances of the MLF.

Orthogonal system of coordinates is recognized to be a powerful tool for analysing multi-generator ac power networks. Choice of orthogonal coordinates in synchronous generators is dictated by its construction; the axis corresponding to the polar axis is conventionally chosen as direct axis, the quadrature being the inter-polar axis leading it by 90°. The choice is somewhat arbitrary for networks. An angle reference is required to be chosen for the entire network; for example, the swing bus in the conventional load flow. Phase angles of all voltages and current phasors are measured with respect to this reference. For joint simulation of generators and networks a coupling matrix is generally employed to relate phasor variables described in the respective reference frames. In Modular approach, we do away with the common reference. For each element we have a different reference. We also do away with the vexed time varying trigonometric transformations. For a passive circuit element i.e., R-X in series, a choice of coordinates is available *naturally*; one aligned with phase of the resistor voltage and the other with that of the inductor voltage. An individual frame of orthogonal coordinates specific to each element is thus possible. Frames of circuit elements will be referred to as orthoframes hereafter.

Hilbert space theory deals with function spaces on which inner products are defined. In orthoframes inner products have a physical significance—*power*—the electrical consumption in the element. This *power* is invariant with respect to choice of reference frames. Space of phasor variables of an element having the inner product—its power loss—constitutes a Hilbert space. With real power *current-squared-r*, is associated a reactive power *current-squared-x* for given current. We

M.V. Hariharan et al., *Modular Load Flow for Restructured Power Systems*, Lecture Notes in Electrical Engineering 374, DOI 10.1007/978-981-10-0497-1_10

show that these two jointly define what is called a Dirac structure for the element. Dirac structures defined on Hilbert spaces have often been employed for analysis of physical systems [1–3] and provide strong theoretical basis in Modular Load Flow. We discuss some preliminaries here.

10.1 Function Spaces with Inner Products

Linear vector spaces qualify as Hilbert spaces as inner product between any two arrow vectors is readily defined. It should however be noted that voltages and currents are not mere geometric arrow vectors. They are constant frequency time functions. Thus they are function vectors, too. The inner product in terms of these functions is defined as,

$$p = \frac{1}{T}\int_{0}^{T} v(t)\, i(t)\, dt \tag{10.1}$$

This is the expression for real power, and in steady state with constant frequency sinusoidal variation of v and i, it can be shown that,

$$p = \frac{1}{T}\int_{0}^{T} v(t)\, i(t)\, dt = V_{rms} I_{rms} \cos\theta \tag{10.2}$$

θ is the phase difference between the two time functions. Due to property (10.2), sinusoidal phasors of voltages and currents can be represented as geometric vectors.

10.2 Hilbert Spaces and Orthoframes

A passive ac circuit element has a characteristic property; in sinusoidal steady state, voltage across reactance x and the voltage across resistance r, are 90° out of phase with each other. A natural choice of orthogonal coordinates for an element is therefore readily available. We define orthoframe Γ_{ei} for element e with respect to generator i (Fig. 10.1),

$$\Gamma_{ei} = \left\{ \overrightarrow{i_{ei}}, \overrightarrow{j_{ei}} \right\} \tag{10.3}$$

Axis $\overrightarrow{i_{ei}}$ is current phasor in element 'e', and is taken as reference. Orientation of axis $\overrightarrow{j_{ei}}$ is perpendicular to it. Only one generator is supposed to be exciting the

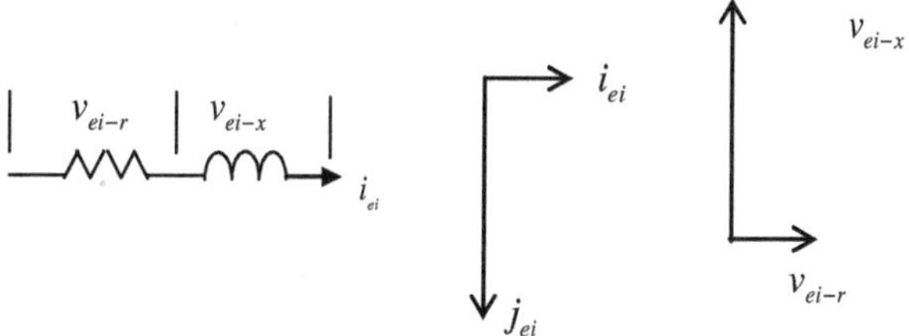

Fig. 10.1 Orthoframe for element *e* with respect to generator *i*

circuit. The two axes of $\overrightarrow{i_{ei}}$ and $\overrightarrow{j_{ei}}$ together form the orthoframe for the element *e*. With $\overrightarrow{i_{ei}}$ as reference, voltage in elements can be determined using ohm's law. Similarly, orthoframe for the generator, is defined by,

$$\Gamma_i \equiv \left\{ \overrightarrow{i_i}, \overrightarrow{j_i} \right\} \tag{10.4}$$

Here i_i is the generator current and taken as reference. Driving point impedance $Z_{ii} = R_{ii} + jX_{ii}$ (the *i*th diagonal element of the bus impedance matrix *Z*) serves as the element impedance. Powers of all generators except the *i*th are set to zero.

Electrical vector space associated with an element is a real Hilbert space and will be denoted by,

$$\mathrm{E}_{ei} \triangleq \{v_{ei}, i_{ei}, p_{ei}\} \tag{10.5}$$

Hilbert space E_{ei} is defined for an element-generator pair. Hence we have two subscripts for variables. Our system description is based on *element variables* and not the *nodal variables.* In order to discover connection between various orthoframes we invoke network theory discussed in the next section.

10.2.1 Relation Between Orthoframes

Consider subsystems with only one generator *i* connected to a network. Multi-terminal description [1] for the network is given by,

$$V = ZI \tag{10.6}$$

Matrix *Z* denotes the bus impedance matrix for the network. With single generator excitation we can write (10.6) as,

$$V_{ivec} = Z_{col-i} I_{ii} \tag{10.7}$$

Vector V_{ivec} contains node voltages as its elements. Let P_i be real power injected by generator i.

$$P_i = I_{ii}^2 R_{ii} \tag{10.8}$$

the current is,

$$I_{ii} = \sqrt{\frac{P_i}{R_{ii}}} \tag{10.9}$$

Magnitude of voltage at the generator bus is given by,

$$|V_{ii}| = |Z_{ii}||I_{ii}| \tag{10.10}$$

And the complex power,

$$I_{ii}^* V_{ii} = I_{ii}^* Z_{ii} I_{ii} = |I_{ii}|^2 (R_{ii} + jX_{ii}) \tag{10.11}$$

$$Q_i = |I_{ii}|^2 X_{ii} \tag{10.12}$$

Bus voltages induced by one and the only generator i are given by,

$$V_{ivec} = Z \begin{bmatrix} 0 \\ 0 \\ . \\ I_{ii} \\ . \\ 0 \end{bmatrix}, \text{ or }, V_{ivec} \triangleq \begin{bmatrix} V_{1i} \\ V_{2i} \\ . \\ V_{nb,i} \end{bmatrix} = \begin{bmatrix} Z_{1i} \\ Z_{2i} \\ . \\ Z_{nb,i} \end{bmatrix} I_{ii}. \tag{10.13}$$

Subsystems for, $i = 1, 2, \ldots, ng$, ng being number of generators, taken together constitute the complete system. Total number of orthoframes in the system is therefore $N_e \times ng$, N_e is the number of elements. We assert that,

$$\frac{r_e}{R_{ii}} = a_{ei}, \text{ (a real constant)} \tag{10.14}$$

Symbol r_e denotes resistance of element e. Linearity of network is assumed. With a single source we have [1],

$$i_{ei} = y_e v_{ei} = y_e \left[A^T V_{ivec}\right]_{\substack{row,\, e \\ col,\, i}} \tag{10.15}$$

Let element e be connected between buses m and n. Then,

$$i_{ei} = y_e [Z_{mi} - Z_{ni}]\, I_{ii} \tag{10.16}$$

The power (loss) in an element is,

$$p_{ei} = |i_{ei}|^2\, r_e \tag{10.17}$$

From linearity of circuit,

$$\frac{|i_{ei}|}{|I_{ii}|} = b_{ei}, \text{(a real constant)} \tag{10.18}$$

Therefore,

$$\frac{|i_{ei}|^2 r_e}{|I_{ii}|^2 R_{ii}} = a_{ei} \times b_{ei}^2 = \varepsilon_{ei}, \text{(a real constant)} \tag{10.19}$$

Expressions $|i_{ei}|^2\, r_e$ and $|I_{ii}|^2 R_{ii}$ represent inner products for element and the generator, and are linearly related via ε_{ei}. Two inner products belong to two Hilbert spaces, one belonging to the element, and the other to the generator. Discovery of this linear relation leads to formulation of MLF. Hilbert spaces admit Dirac structures [2].

10.3 Dirac Structures

Linear vector spaces associated with an electrical network are endowed with interesting structural properties. Examples and analogies will be used to gain an intuitive understanding of the idea. Mathematical details can be found in the references [2–4]. First we define and explain a few terms.

10.3.1 Dual Spaces

Let i and v be two linear vector spaces of currents and voltages. They are called dual of each other if there is a linear functional relationship between the two spaces. For example, in electrical circuits, a linear functional relationship exists between voltage v and the current i in a circuit element through its impedance. Voltage and current spaces of an element are therefore dual of each other.

10.3.2 Duality Product

Duality product is the inner product denoted by $\langle v \mid i \rangle$. It has a magnitude $VI \cos \varphi$ for sinusoidal voltages. Vectors of spaces i and v are restricted vector spaces of ordered pairs of element currents and voltages in the network.

10.3.3 Bilinear Form

There exists an inner product denoted by $\langle v \mid i \rangle$, which in d-q variables is denoted by $v_d i_d + v_q i_q$. We identify v_d, v_q with v^1, v^2 and i_d, i_q with i^1, i^2, so that the bilinear form, $\langle\langle (v^1, i^2), (v^2, i^1) \rangle\rangle$ is same as $v_d i_q + v_q i_d$.

Inner product is recognized as active or, real power and the associated bilinear form as reactive power. Please note that magnitudes of inner product and bilinear forms are invariant with respect to the choice of the d-q frame.

10.3.4 Orthogonal Complement

Let's choose a frame comprising the d-q axes for a circuit element aligned along resistor voltage and inductor voltage, respectively (Fig. 10.2).

Let v_d and i_q define a space D. Then, the space defined by v_q and i_d is orthogonal complement of D and is denoted by $D^{\perp}$. From observation,

$$D = D^{\perp} \tag{10.20}$$

This condition defines a Dirac structure with v_{ei} *and* i_{ei} spaces. The context is a little different from that in [4].

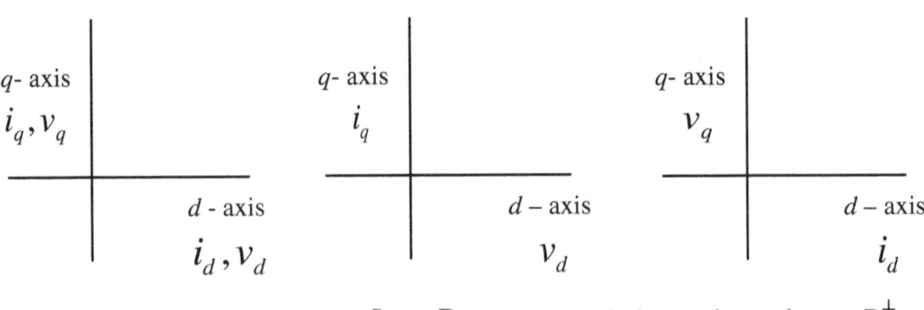

Fig. 10.2 Dirac structure

10.3.5 Dirac Structure: Definition

Dirac structure is a vector space D such that, D = $D^{\perp}$, where $D^{\perp}$ represents the orthogonal complement of D (Def. 2.2 [2]).

10.4 Dirac Structure for Generators

Consider only *one* generator injecting power P_i into a network consisting of transmission lines and loads. Let the driving point impedance at the generator port be given by $Z_{ii} = R_{ii} + jX_{ii}$. Let V_{ii} *and* I_{ii} be the voltage and current at this port. Inner product in terms of rms values of voltage V_{ii} and current I_{ii} is, $V_{ii}I_{ii}\cos\varphi_i$. For an arbitrary choice of frame, the *d* and *q* axis, we write,

$$P_i = V_{ii}I_{ii}\cos\varphi_{ii} = V_{dd}I_{dd} + V_{qq}I_{qq} \tag{10.21}$$

We shall assign voltage across R_{ii} as *d* axis and that across X_{ii} as *q* axis to simplify this expression further.

The bilinear form associated with (10.21) is,

$$Q_i = V_{dd}I_{qq} + V_{qq}I_{dd} = V_{ii}I_{ii}\sin\varphi_{ii} \tag{10.22}$$

10.5 Geometric Interpretation

Generator powers are independent of one another. Inner products *in the element* are independent as these are its fractions of independent generator powers. On each axis therefore we can plot Hilbert spaces for these powers. These constituent Hilbert spaces in an element *get decoupled by virtue of independence of the generators to which they belong.* We illustrate these in Fig. 10.3.

The circles show the Hilbert spaces of generator 1 and 2 in element *e*. Decoupling shown in Fig. 10.3 can be easily extended to more generators.

10.5.1 Electrical Variables in Hilbert Spaces

Electrical variables of Hilbert space of generator 1 and generator 2, *in element e* (i.e., Hilbert spaces of the fractional powers of generators) are shown in Fig. 10.4. Voltages v_{e1} *and* v_{e2} are two generator voltage contributions in element *e*. v_{e1} is drawn in the plane of generator 1 and v_{e2}, in the plane of generator 2. We see that v_{e1} and v_{e2} are also orthogonal.

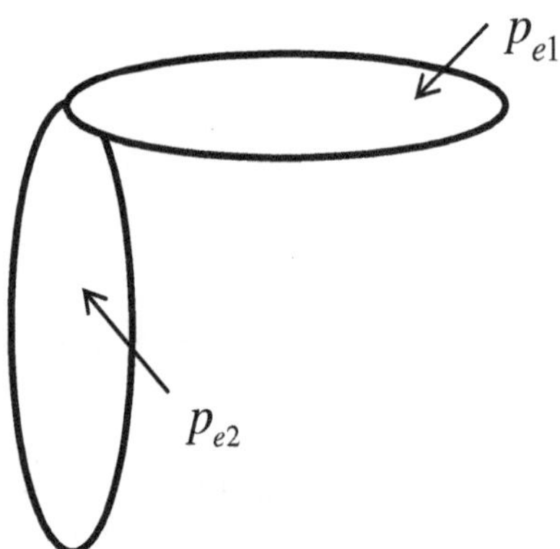

Fig. 10.3 Power spaces in an element

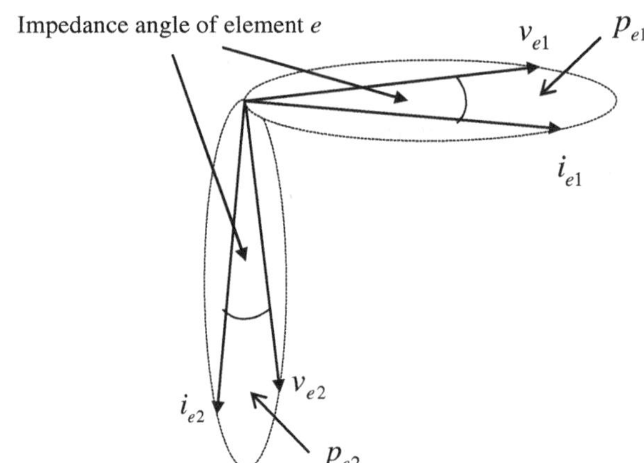

Fig. 10.4 Electrical variables of generators in element *e*

10.5.2 Superposition of Dirac Structures

Dirac structures in element *e* can be superposed. This idea is akin to the *many parallel world interpretation* in quantum theory wherein *Pauli's exclusion principle* applies to like particles which can exist independently without interacting and can be superposed. The inner products, being scalars, can be therefore summed directly. To obtain v_e from v_{e1} and v_{e2}, we must extract desired vectors from Fig. 10.4 as shown in Fig. 10.5. Numerical values are used from the example that follows. We see that voltage v_e and current i_e are the Pythagorean sums,

$$v_e^2 = \sum_i v_{ei}^2 \tag{10.23}$$

$$i_e^2 = \sum_i i_{ei}^2 \tag{10.24}$$

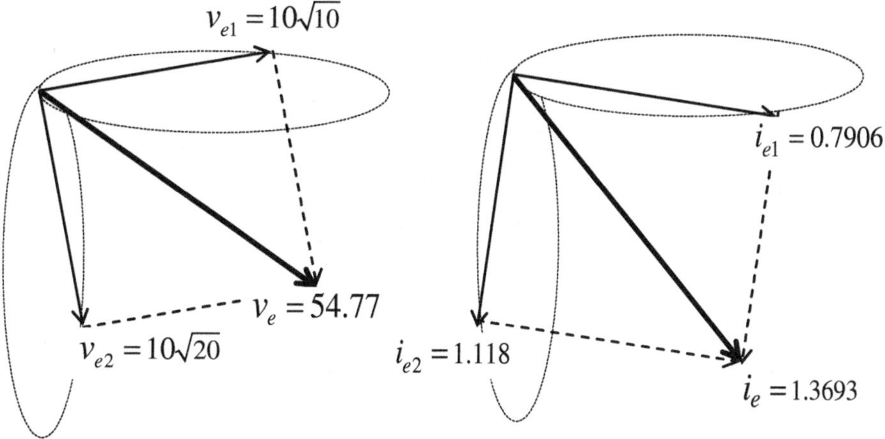

Fig. 10.5 Superposition of electrical variables in 40 Ω element

10.5.3 Example

Consider the example from Sect. 3.3.1. With generator 1, we had calculated for element 1 (40 Ω),

$$v_{11}(=v_{40,1}) = 10\sqrt{10} \quad and \quad i_{11}(=i_{40,1}) = 0.7906$$
$$v_{12}(=v_{40,2}) = 10\sqrt{20} \quad and \quad i_{12}(=i_{40,2}) = 1.118$$

The Pythagorean sum then gives the net voltage and current in the element of 40 Ω. That is,

$$v_1^2 = \sum_i v_{1i}^2 = \left(10\sqrt{10}\right)^2 + \left(10\sqrt{20}\right)^2$$
$$= 3000$$

Or,

$$v_1 = \sqrt{3000} = 54.77$$

This agrees with conventional calculation, $v_1 = \sqrt{P_{total} * R_{net}} = \sqrt{300 * 10} = 54.77$

$$i_1^2 = \sum_i i_{1i}^2 = \left(\frac{1}{40} * (10\sqrt{10})\right)^2 + \left(\frac{1}{40} * (10\sqrt{20})\right)^2 \text{Or, } i_1 = \sqrt{1.875} = 1.3693$$
$$= 0.7906^2 + 1.118^2 = 1.875$$

Similarly, in the element 40/3 Ω,

$$v_{21}(=v_{40/3,2}) = 10\sqrt{20} = 44.7214 \quad and \quad i_{21}(=i_{40/3,2}) = \frac{44.7214}{40} = 1.118$$

$$v_{22}(=v_{40/3,2}) = 10\sqrt{20} = 44.7214 \quad and \quad i_{22}(=i_{40/3,2}) = \frac{44.7214}{(40/3)} = 3.3541$$

$$\begin{aligned} v_2^2 &= \sum_i v_{1i}^2 = \left(10\sqrt{10}\right)^2 + \left(10\sqrt{20}\right)^2 \text{ Or,} \\ &= 3000 \\ v_2 &= \sqrt{3000} = 54.77 \\ i_2^2 &= \sum_i i_{1i}^2 = \left(\frac{1}{40} * 10\sqrt{20}\right)^2 + \left(\frac{1}{(40/3)} 10\sqrt{20}\right)^2 \\ &= 1.118^2 + 3.3541^2 = 12.50 \\ \text{Or, } i_2 &= \sqrt{12.50} = 3.5355 \end{aligned}$$

Numerical values for element −1 (40 Ω) are shown on Fig. 10.5. All final values obtained above can be verified with conventional calculations.

10.6 Summary

Hilbert spaces and Dirac structures associated with power networks provide interesting area for analytical research. Electrical variables of an element are shown to be Pythagorean sums of the individual generator contributions in the element. Individual power contributions in elements, on the other hand, are directly summable. Mismatches, though small, do appear in power balance when all generators are considered together because of interactions among generators which are small in power networks. In large systems with many generators total power consumption in elements is found to be slightly less than the injected turbine powers. The difference can be interpreted as power lost in *tying up* individual Dirac structures. For power networks, this small power loss can generally be neglected. Concept of entanglement in quantum theory seems relevant here. Material given in this chapter should be considered as an initial idea not fully developed. It should open possibilities for further research in multi-generator power systems.

References

1. M.A. Pai, *Computer Techniques in Power System Analysis* (Tata McGraw-Hill, New Delhi, 1979)
2. G. Golo, O.V. Iftime, H.J. Zwart, A.J. Van Der Schaft, Tools for analysis of Dirac Structures on Hilbert Spaces, *Memorandum No. 1729*, Department of Mathematics, University of Twente, Netherland (2004)

3. B.M. Maschke, A.J. van der Schaft, Interconnection of systems: the network paradigm. in *Proceedings of the 35th Conference on Decision and Control, Kobe, Japan* (1996)
4. R. Pasumarthy, Ph.D. Thesis, On analysis and control of interconnected finite-and infinite dimensional physical systems, Dutch Institute of Systems and Control, 1978

Zeitfracht Medien GmbH
Ferdinand-Jühlke-Straße 7
99095 Erfurt, Deutschland
produktsicherheit@kolibri360.de